SMP 11-16

Book G+

CAMBRIDGE
UNIVERSITY PRESS

Published by the Press Syndicate of the University of Cambridge
The Pitt Building, Trumpington Street, Cambridge CB2 1RP
40 West 20th Street, New York, NY 10011–4211, USA
10 Stamford Road, Oakleigh, Melbourne 3166, Australia

First published 1994

Produced by Gecko Limited, Bicester, Oxon.

Printed in Great Britain by Scotprint, Musselburgh, Scotland

A catalogue record for this book is available from the British Library

ISBN 0 521 45795 5 paperback

Cover: Art Wolfe/Zefa

Contents

	Making tens	4
1	Reflection	5
	Jigsaw	9
	Perimeters	10
2	Flow diagrams	11
	Look, think, check	16
3	There's always more than one way . . .	17
	Estimating length	24
	Two hundred and one	25
4	What's the difference?	26
	Estimating capacity	32
5	Shapes and sequences	33
	Boxes	38
6	Leftovers	41
	Flying away from it all	47
7	The results prove it!	50
	Angles and shapes	55
8	Negative numbers	59
	Divide and rule	66
	Reviews	68

Making tens

This is a game for two.
You need five 10-sided dice.

- Players take it in turns to throw the five dice.
- From the numbers showing players group the dice so that the numbers add up to 10.
- Points are scored like this:

10 with two dice	two points
10 with three dice	three points
10 with four dice	four points
10 with five dice	five points

For example:

The total points scored for this throw is five.

- The first player to reach 20 points wins.

discussion point

How can making tens help you quickly work out calculations like:
$3 + 6 + 4 + 7 + 3$?

1 Reflection

A Reflection symmetry

This is a chapter for two people to do together.

This traffic sign has reflection symmetry,

but the real car does not have reflection symmetry.

A1 Which of these have reflection symmetry?
It may help you to look at the real-life objects.

(a)

(b)

(c)

(d)

A2 Make a list of five objects which have reflection symmetry and five objects which do not have reflection symmetry.

Is it easier to find objects with reflection symmetry or without reflection symmetry? Check each other's lists.

This pattern is symmetrical.
The dotted line is called a **line of reflection symmetry**.

(Some people miss the word 'reflection' out and just say 'symmetry'.)

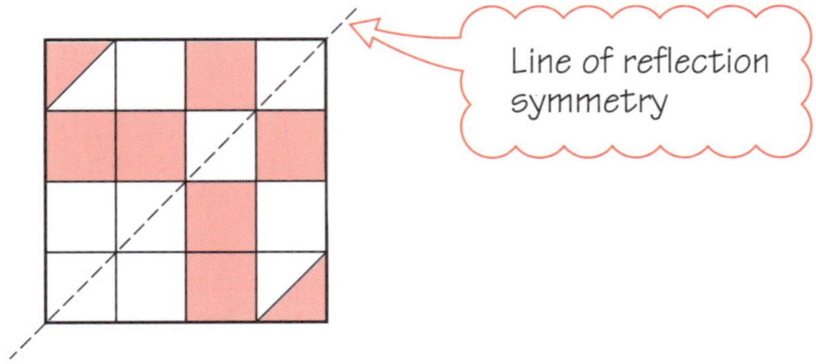

A3 Copy these diagrams onto squared paper.

Complete the shading so that the dotted lines are lines of reflection symmetry.

(a)

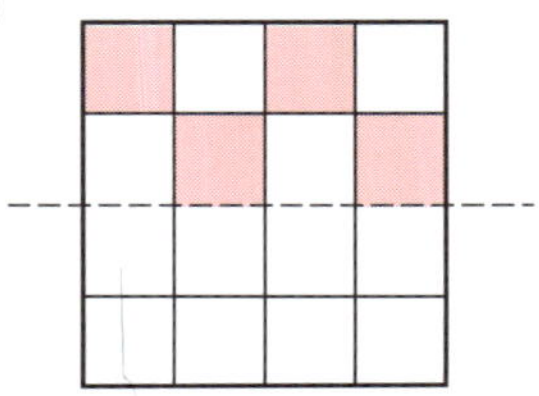

(b)

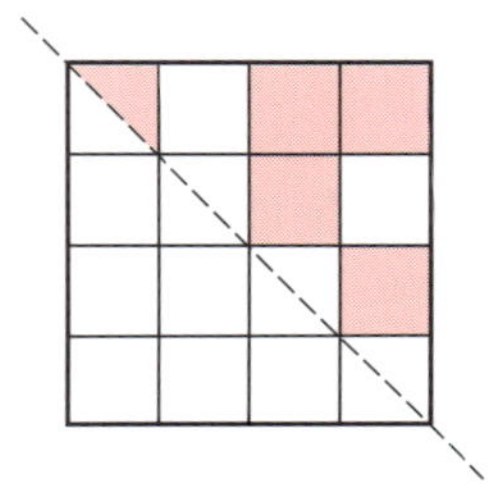

(c)

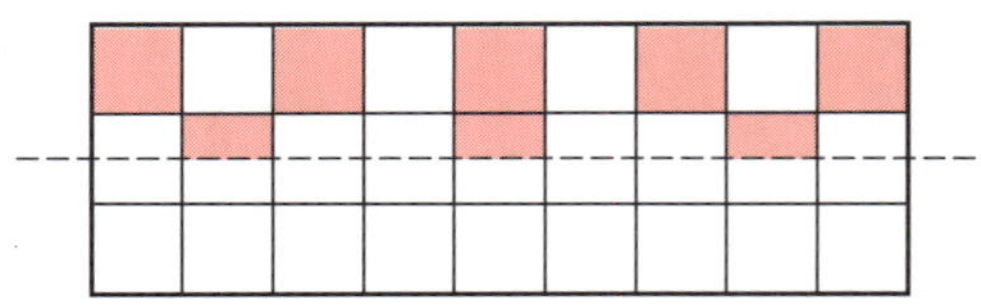

A4 Make an interesting pattern which has a line of reflection symmetry.

discussion point

How would you convince someone that this dotted line is **not** a line of reflection symmetry for the shape?

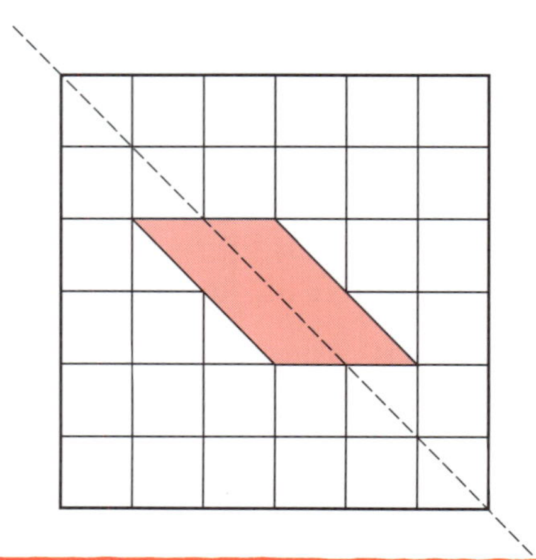

A5 Look at these ancient patterns together.
Decide which ones have reflection symmetry.

(a)

(b)

(c)

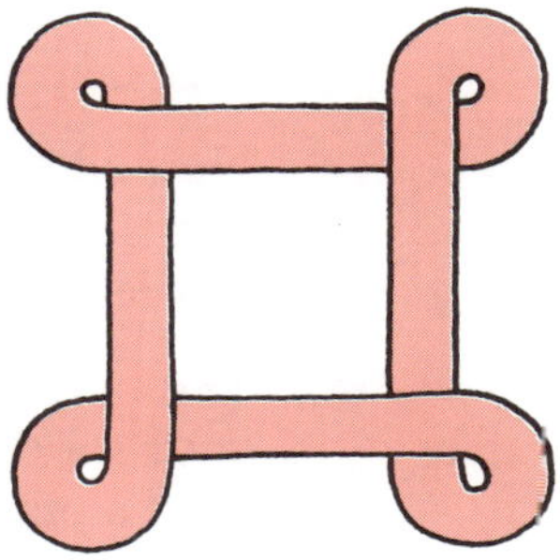
(d)

B Solid shapes

You need 16 multilink cubes of one colour and 6 of a different colour.

1

You need 8 multilink cubes, all the same colour. Make this solid shape.

2

Now split the solid into two parts like this.

3

Put a mirror between them.

Now lift the mirror up and down slowly. The solid shape does not change. This is because one side is a reflection of the other.

The solid shape has reflection symmetry. The mirror is a **plane of symmetry**.

B1 You need 12 cubes, all the same colour.

Use 6 of the cubes to make this shape.

Now add the other 6 cubes to make a solid shape which has reflection symmetry.

Use a mirror to check that your solid shape does have a plane of reflection symmetry.

B2 You need 16 cubes, all the same colour.

Use 8 of the cubes to make this solid shape.

Now add the rest of the cubes to make a solid shape which has a plane of reflection symmetry.

Use a mirror to check that it does.

B3 You need 8 cubes of one colour and 6 of another.

Use 7 cubes to make this solid shape.

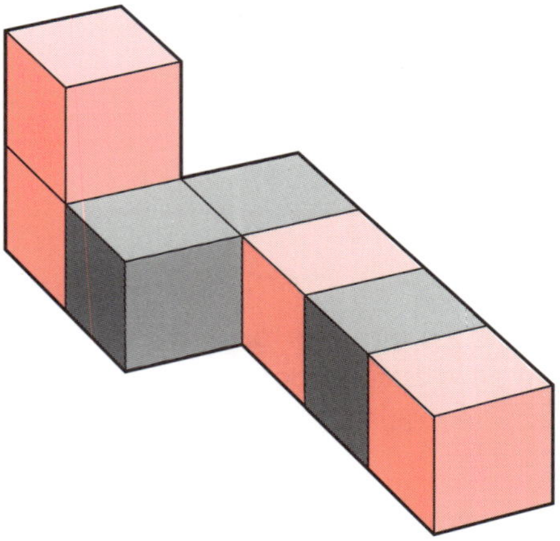

Add the rest of the cubes to make a solid shape which has a plane of reflection symmetry.

Jigsaw

You need worksheet G+ 1 and some tracing paper.

This is a novelty jigsaw puzzle.

It is double sided and **both** the sides are plain.
In other words each piece could fit either side up.

Shapes which match each other exactly are called **congruent**.

This shape 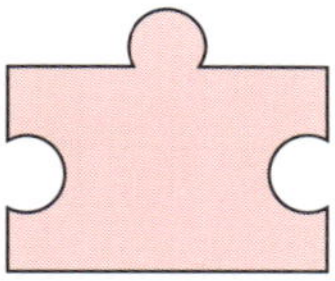 is congruent to this shape 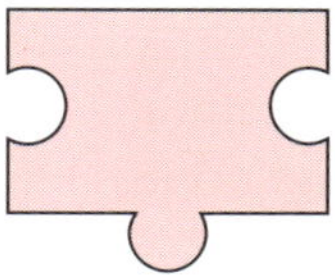

Use tracing paper to check that this is true.

These two shapes are also congruent – convince yourself by using tracing paper. (You can turn the tracing paper over.)

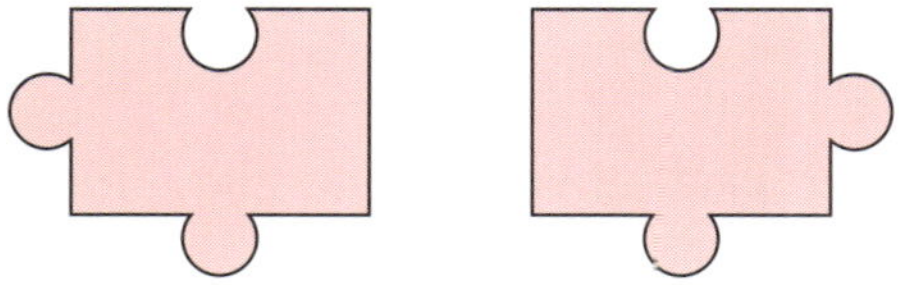

On worksheet G+ 1 are the 16 pieces which make up the jigsaw.

How many sets of pieces can you find which are congruent?
For example, the pieces labelled (b) and (e) are congruent.
(Don't forget to use tracing paper to make sure!)

Challenge

Try to do the jigsaw. (You will need scissors.)

Perimeters

1 Suppose you have 12 tiles. Each tile is 1 cm by 1 cm.
You can arrange them in a rectangle in different ways.

Here is one way – a rectangle 4 cm by 3 cm.
(We sometimes leave out the cm and call this a 4 by 3 rectangle.)

Draw as many different rectangles as you can using all 12 tiles.

2 The **perimeter** of the rectangle is the distance all the way round the outside.

For the rectangle above, the perimeter is $3 + 4 + 3 + 4 = 14$ cm.
Copy this table.

Size of rectangle	Perimeter
4 by 3	14 cm

Fill in the size and perimeter for each of your rectangles.
What size has the **largest** perimeter?
What size has the **smallest** perimeter?

3 Now suppose you have 20 tiles.
How many different rectangles can you make with them?

Record their size and perimeter in a table like the one above.

Which size rectangle has the largest perimeter?
Which has the smallest perimeter?

4 Try some different numbers of tiles for yourselves.
Which size rectangle has the largest perimeter?
Which size has the smallest?

2 Flow diagrams

A A puzzle and a game

A1 Here is a puzzle for two to work on together.

Three women and two girls need to cross a dangerous river.

There is only one small boat available.
It will hold either two girls or one woman.

How can they all cross the river safely?
(You may need some counters.)

Make sure it doesn't add up!

This is a game for two people.
You need a set of 1 to 9 number cards and two pieces of paper.

- Players take it in turns to put down a number card on one of the sheets of paper. You cannot put down a card when it equals two of the cards already on the sheet, added together.

  So if 3 and 4 are on one sheet of paper you cannot put down the 7 on the same sheet. If 1 2 and 5 are on the other sheet then you cannot put down the 7 on here either.
- You must put the cards down in order starting with 1.
- You can choose which sheet to put the card on.
- When you cannot put down a card you lose the game and your opponent gets one point. The first player to score five points wins.

Play 'Make sure it doesn't add up!' yourselves.
Is there a method (we sometimes say **strategy**) of making sure you win?

Here is the solution to the puzzle in question A1.

A **flow diagram** will also show how to solve the puzzle. Just follow the flow of the arrows.

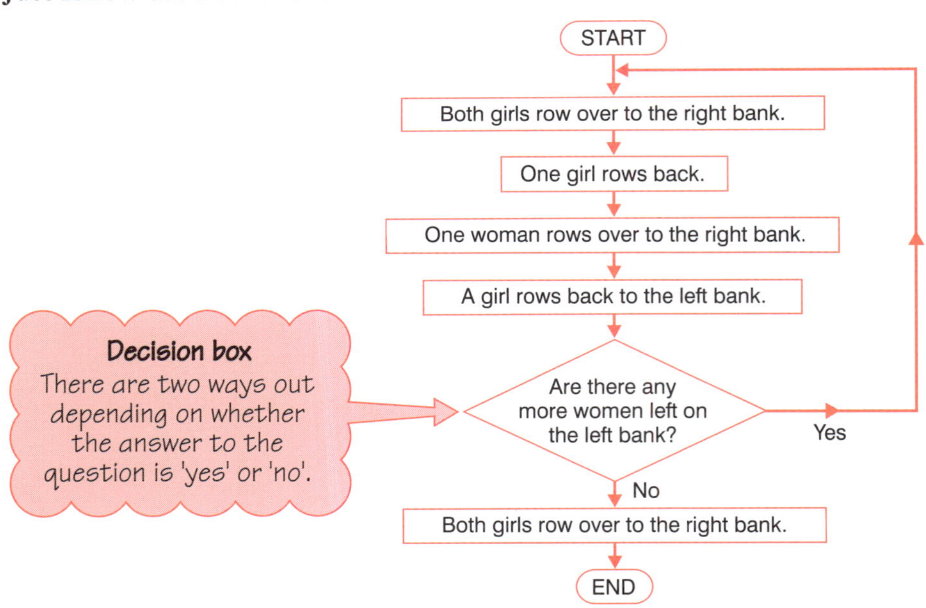

A2 Check that the flow diagram works by using it to 'help' all the people across the river.

A3 Use this flow diagram to help you play 'Make sure it doesn't add up!' Make sure your opponent cannot see it!

Check that you always win if you follow this flow diagram. You need to go second. Call one sheet of paper A and the other B.

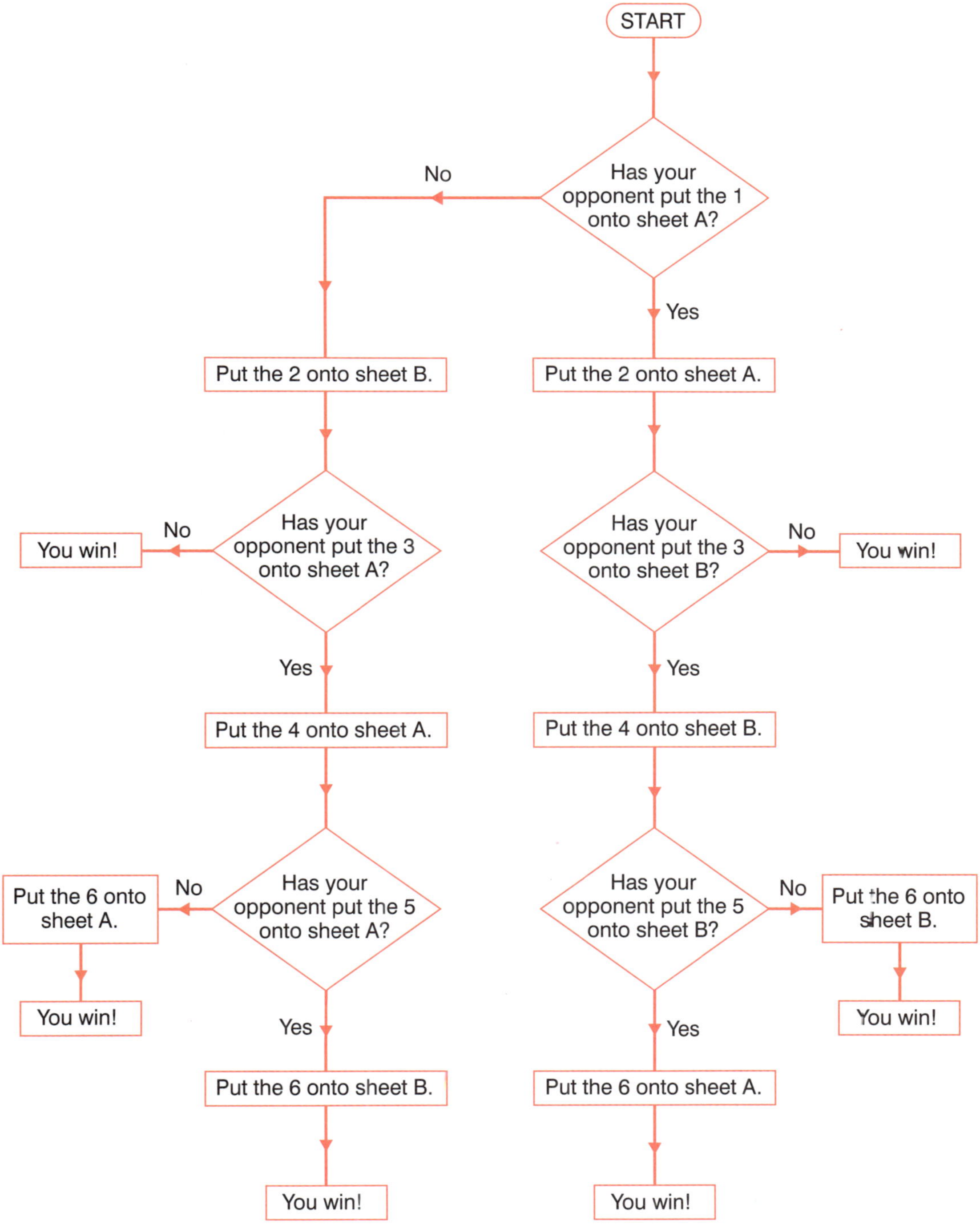

B Using flow diagrams

Many people learn first aid. A group of doctors in Europe agreed on the best way for these people to deal with a casualty. They wrote a flow diagram to show what to do.

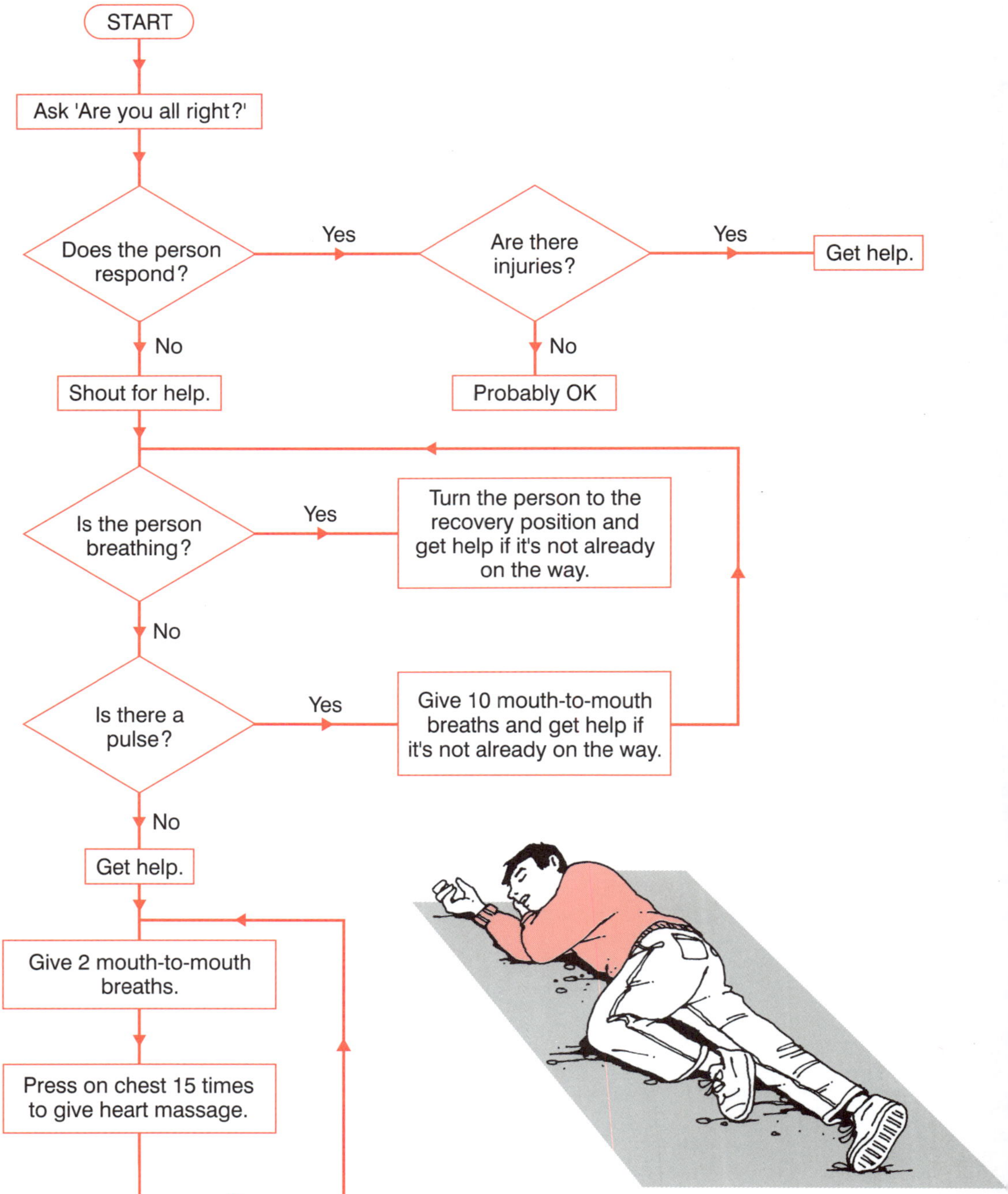

B1 Imagine you find someone lying on the ground.
Use the flow diagram to work out what you should do if:

(a) the person responds when you talk to them and has no sign of any injuries

(b) the person does not respond, has a pulse but is not breathing

(c) the person does not respond but is breathing

(d) the person does not respond, has no pulse and is not breathing.

B2 You see someone giving heart massage at a road accident.
What can you say about the casualty?

B3 Flow diagrams can be used to sort or **classify** objects, for example trees, insects or fish.

Here is a flow diagram which sorts the whole numbers from 1 to 20.
The shaded boxes show the sorted numbers.

Try to work out what question should go into the two decision boxes.

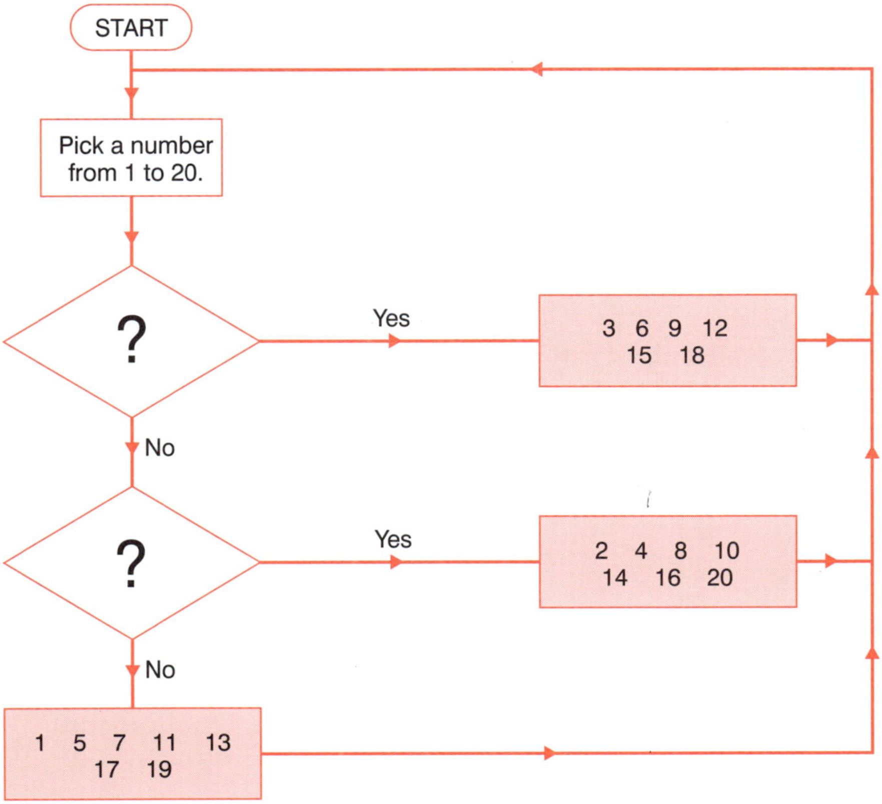

Look, think, check

1 Here are some number patterns. Copy and complete the first four lines. You may use your calculator for this.

Now look at the pattern of numbers and **without** a calculator complete the next three lines.

Use a calculator to check these answers.

(a)

12 x 11 = 132
13 x 11 = 143
14 x 11 =
15 x 11 =
16 x 11 =
17 x 11 =
   x 11 =

(b)

9 x 3 = 27
99 x 3 = 297
999 x 3 =
9999 x 3 =
99999 x 3 =
999999 x 3 =
9999999 x 3 =

(c)

34 x 3 = 102
34 x 6 = 204
34 x 9 =
34 x 12 =
34 x 15 =
34 x   =
34 x   =

2 Copy and continue these patterns of numbers. Check that how you think the sequence carries on is correct by using a calculator. How far is it possible to continue these sequences?

(a)
$$\begin{array}{l} 1 + 2 - 3 = 0 \\ 2 + 3 - 4 = 1 \\ 3 + 4 - 5 = \dots \\ \vdots \end{array}$$

(b)
$$\begin{array}{ll} 1 + 1 & = 2 \times 1 \\ 12 + 21 & = 3 \times 11 \\ 123 + 321 & = \dots \times \dots \\ \vdots & \end{array}$$

(c)
$$\begin{array}{lll} 9 + 1 & = & \dots \\ 98 + 12 & = & \dots \\ 987 + 123 & = & \dots \\ \vdots & & \end{array}$$

(d)
$$\begin{array}{lll} 9 - 1 & = & 8 \\ 98 - 21 & = & 77 \\ 987 - 321 & = & 666 \\ \vdots & & \end{array}$$

3 Copy and complete these patterns.
You may use your calculator if you wish.

(a)
$$\begin{array}{lcl} 32 \times 27 & = & 864 \\ 16 \times 54 & = & 864 \\ 8 \times 108 & = & \dots \\ \dots \times 216 & = & 864 \\ \dots \times 432 & = & 864 \\ 1 \times \dots & = & 864 \end{array}$$

(b)
$$\begin{array}{lcl} 64 \times 15 & = & 960 \\ 32 \times 30 & = & 960 \\ 16 \times \dots & = & 960 \\ 8 \times \dots & = & 960 \\ \dots \times 240 & = & 960 \\ \dots \times 480 & = & 960 \\ 1 \times 960 & = & \dots \end{array}$$

Look at the first column of the multiplications. What do you notice?
What do you notice about the second column of the multiplications?

3 There's always more than one way ...

This chapter is about using pencil and paper to work out multiplications. Only use a calculator to check your answers.

discussion point

Look at each of these and check for yourselves that they work. Try to decide **why** they work.

A Spots and dots

Safiq and Lindsay need to find out how many trees there are in an orchard. They are working from a plan. All the trees are planted in rows.

discussion point

Safiq suggests that they just count all the trees.
But Lindsay says there must be a quicker way.
Can you think of any other ways of finding how many trees there are?

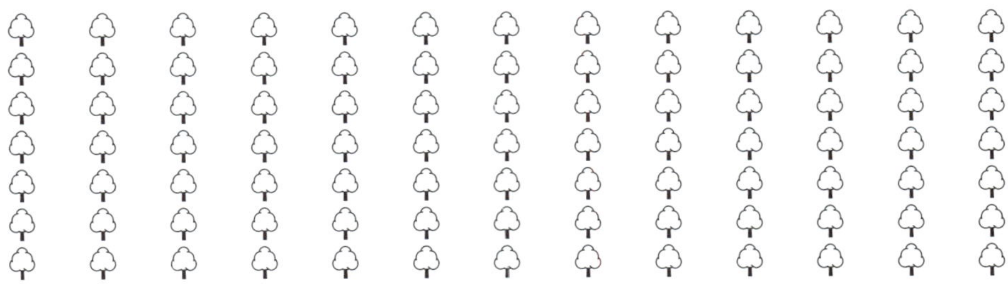

Lindsay suggests splitting up the trees into smaller groups like these.
How would this help them find the total number of trees?

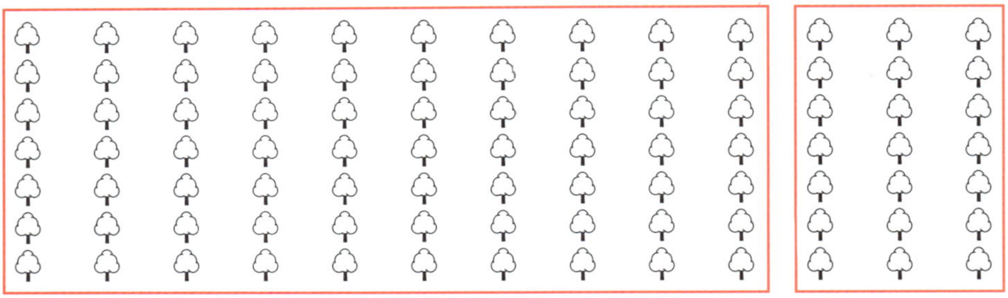

A1 Try to work out how many trees there are **without** using a calculator.

To calculate the number of spots in this block you need to work out 16×7.
16×7 can be split into simpler calculations.
For example $16 \times 7 = (10 \times 7) + (6 \times 7)$

$$= 70 + 42$$
$$= 112$$

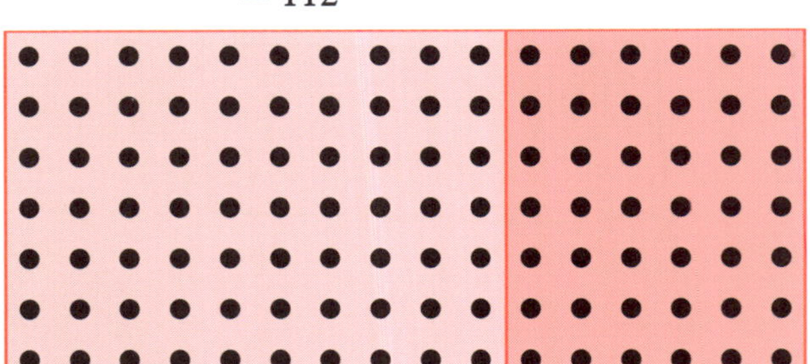

Or, $16 \times 7 = (10 \times 5) + (10 \times 2) + (6 \times 5) + (6 \times 2)$ as shown below.
$= 50 + 20 + 30 + 12$
$= 112$

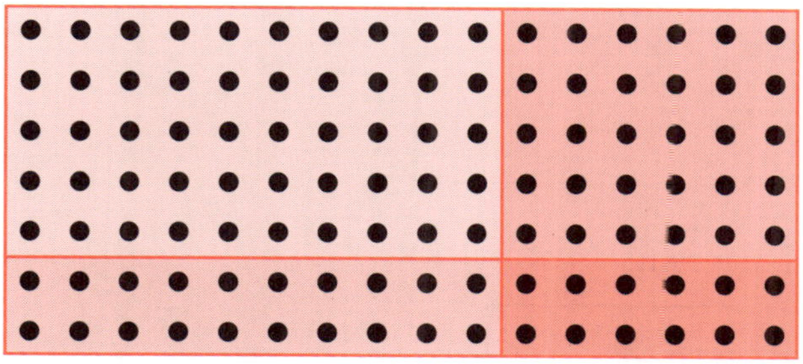

A2–A4 These questions are on worksheets G+ 2 and G+ 3.

A5 On a piece of square spotty paper draw round a block of 14 spots by 7 spots. Split this block up into smaller blocks which will help you work out the value of 14×7.

Use a calculator to check your answer.

To save counting spots, it is easier to work with areas.

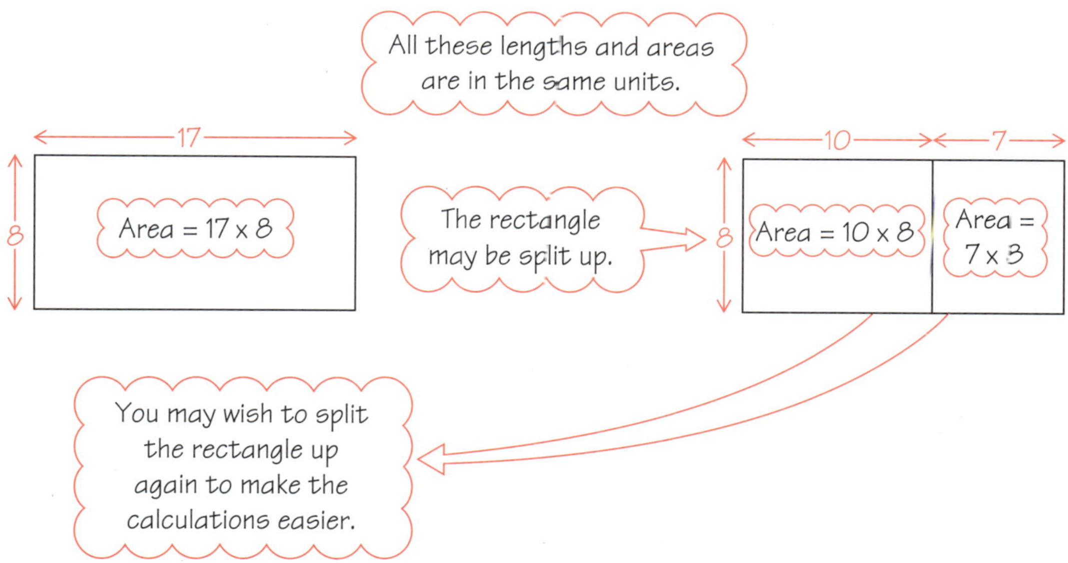

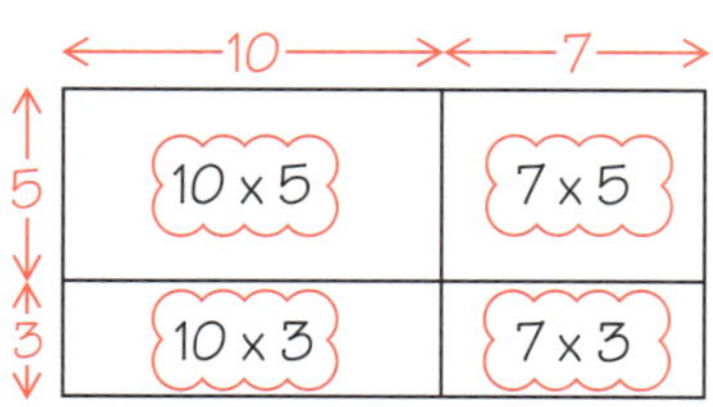

17 x 8 is the same as:
$(10 \times 8) + (7 \times 8) = 80 + 56$
$= 136$
or
$(10 \times 5) + (7 \times 5) + (10 \times 3) + (7 \times 3)$
$= 50 + 35 + 30 + 21$
$= 136$

A6 Find the area of this rectangle – without using a calculator.

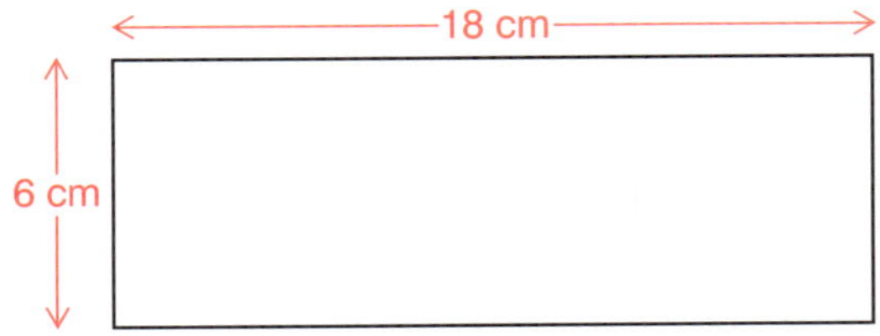

You don't always have to use a diagram!

Twenty-one lots of eight

Splitting 21 up into 10 + 10 + 1

21×8
$= 10 \times 8 + 10 \times 8 + 1 \times 8$
$= 80 + 80 + 8$
$= 168$

Or, splitting 21 up into 20 + 1

$21 \times 8 = 20 \times 8 + 1 \times 8$
$= 160 + 8$
$= 168$

A7 **You may use a calculator for this question!**
Investigate this pattern of multiplications.

$4 \times 6 = 24$
$6 \times 4 = 24$
$4 \times 60 = 240$
$6 \times 40 =$
$4 \times 600 =$
$6 \times 400 =$
and so on …

Make a note of anything interesting you find.

A8 Use the results you found in question A7 to work these out.

(a) 7×30 (b) 40×9 (c) 30×9 (d) 80×5

(e) 300×9

Check your answers with a calculator.

A9 Use any pencil-and-paper method you wish to answer these.

(a) 33×7 (b) 9×43 (c) 86×5 (d) 6×83

B The window method

This is a very old method. It came from India many years ago.

It is called the window method because it uses a grid shaped like an old window.

Work through this with a partner.

1

Here is how the method works for 45×8.

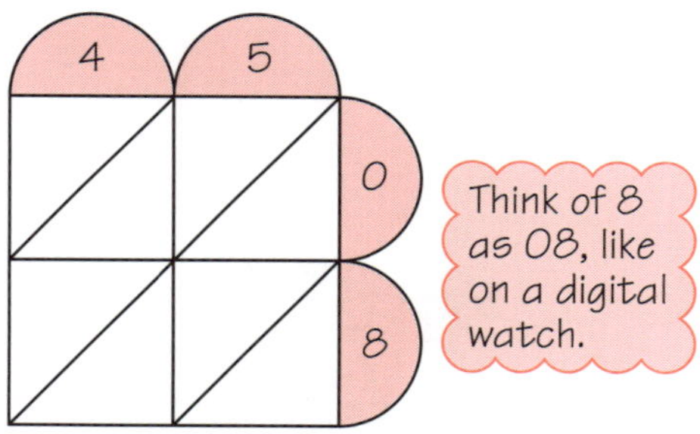

2

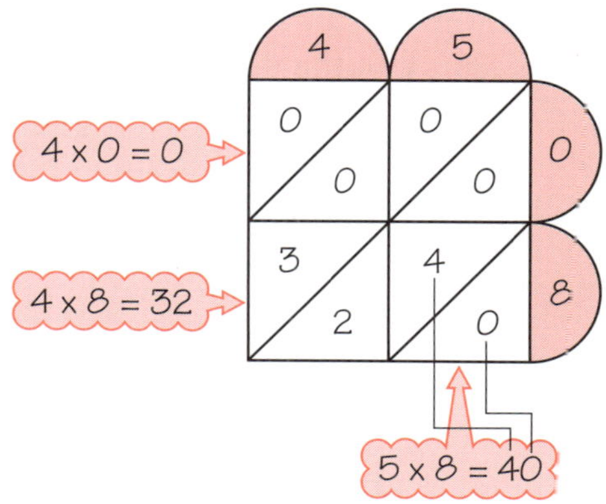

3

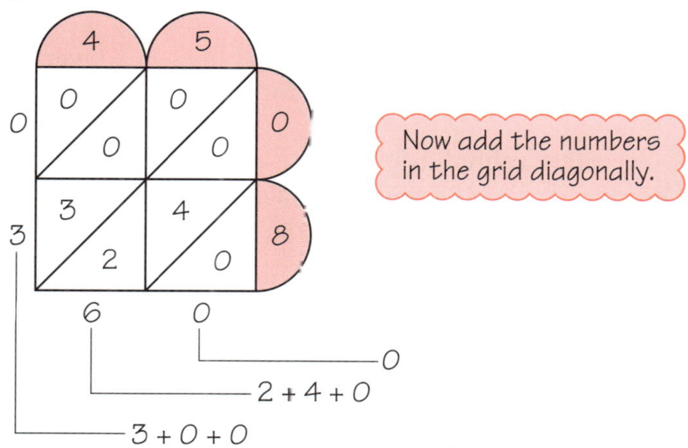

So 45 x 8 = 360

B1 Use the window method to calculate these.

(a) 54×8 (b) 43×6 (c) 7×26

Check your answers with a calculator.

Look carefully at this example.

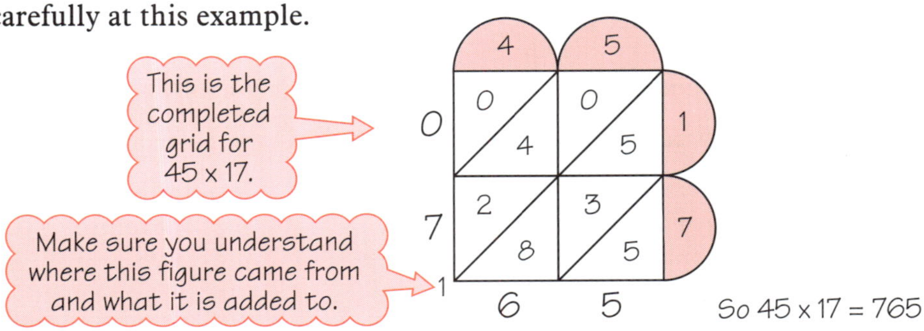

B2 Use the window method to calculate these.

(a) 69×8 (b) 65×7 (c) 9×47 (d) 65×23

B3 Try to work out the answer to 83×97.
Use a calculator to check your answer.

B4

A supermarket trolley will hold 15 boxes in a layer.
How many boxes will there be in 7 layers?

B5 A secondary school has 25 classes.
There are exactly 28 pupils in each class.
How many pupils are there altogether?

B6

School chairs cost £18 each.
How much would it cost to buy new chairs for a class of 33?

C Using multiplication

Use whatever pencil-and-paper method you wish to answer these.

C1

The nine o'clock news lasts 25 minutes each night.
It is on from Monday to Friday.
How many minutes is this each week?

C2 Every hour there are 11 thousand more people in the world!
How many thousands is this in 24 hours?

C3 Isha saves £6 a week. How much will she save in a year?
(There are 52 weeks in a year.)

C4 Pupils at Archway School go to school for 37 weeks of the year.
Every week they have 5 maths lessons.
Each lesson lasts for 35 minutes.

(a) How many maths lessons do the pupils have in a year?

(b) How many minutes of maths do they have in a week?

C5 Lee's car will go for 47 miles on 1 gallon of petrol when its average speed is 55 m.p.h. How far will it go on 9 gallons at 55 m.p.h?

C6 It costs Graham 78p a day to travel to and from school.
How much will it cost him to travel in a week?

C7

A hospital block has 6 floors. There are 82 beds on each floor.
How many beds are there altogether?

Estimating length

These activities are for one or two people.

1 Estimate the actual height of this tree.

Describe how you worked out your estimate.
Compare answers and methods with your neighbour.

Estimating is **not** the same as guessing.
When we estimate the height of something we usually compare it with a height or length we know.

2 Explain why it is impossible to estimate the actual height of this flower.

You can compare the lengths and heights of objects in your room to the length of a metre rule.

3 Estimate these:

(a) the height of the door

(b) the height of the room

(c) the height of a cupboard

(d) the length of the room

(e) the width of the room.

Don't forget to put the units in your estimates!

4 Measure the distances you estimated in question 3.
Which was your best estimate? Which was your worst?

5 Now that you know the measurements of your room, estimate these:

(a) the length of the corridor

(b) the height of the school.

6 Estimate the distance to the school office.

Show your answers to your teacher.

Two hundred and one

This is a game for two.
You need two ordinary 6-sided dice.

Players take it in turn to throw the two dice.
The score is the two numbers showing on the dice multiplied together.

Write this calculation and the answer like this:

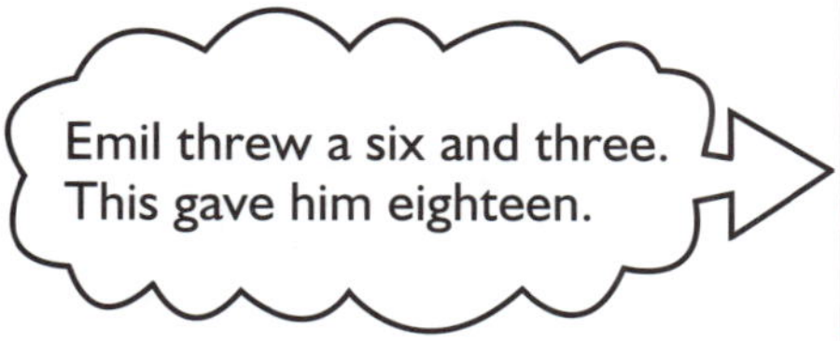

Dice calculation	Score	Total so far
6 × 3	18	18
5 × 5	25	43
5 × 6	30	

On the next throw he got two fives. He multiplied these making twenty-five and added this to the eighteen.
This gave forty-three.

The object of the game is to get as near as possible to 201 without going over it.
Players may 'stick' at any time they wish.

No calculators are allowed!

When you have played the game a few times:

- play it with two 0–9 dice to get close to 301
- allow a player to add, subtract or multiply the two numbers showing on the dice
- make up a similar game for yourselves and test it on your friends

4 What's the difference?

This is a chapter for two people

A Better, worse or about the same?

Two friends, Emma and Karen, are arguing about whose class had the better attendance over the last week.

The better class would win a 'no uniform' day!

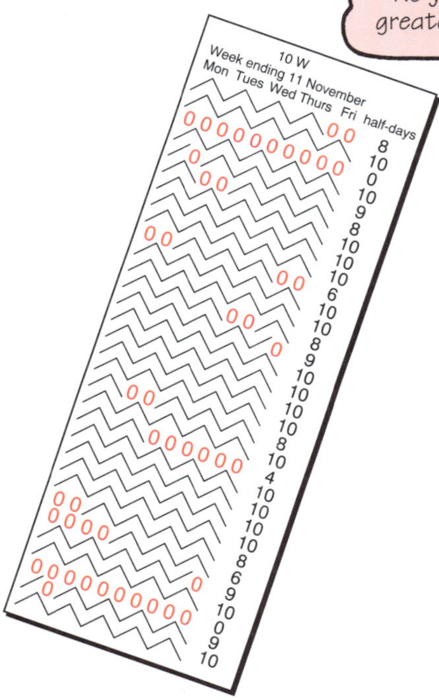

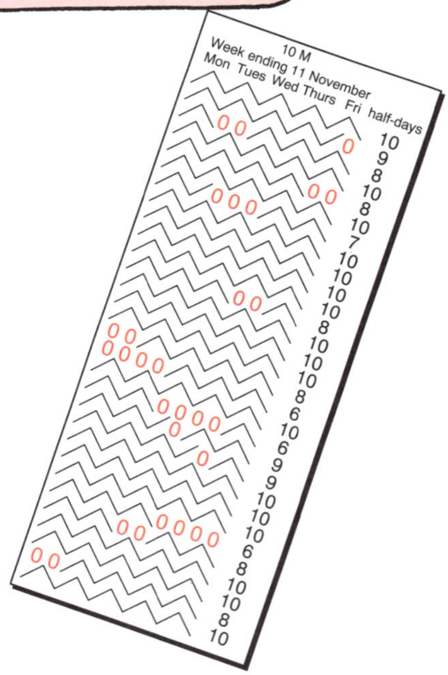

A1 One of you use the figures for 10M, the other for 10W. Try to convince the other person that your class has the better attendance record.

Write down your reasons and show them to your teacher.

A2 Karen says that adding together the number of half-day attendances for each class and using this total is a fair method.

Is she right? Talk this through together.

Karen works out the total attendances for each class.

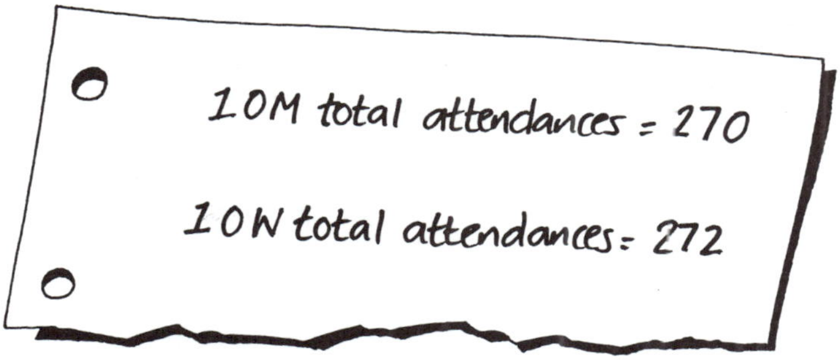

But Emma says …

A3 How many students are there in 10W?

One way to allow for the different number in each class is to find the **average** (or **mean**) number of attendances.

10M had a total of 270 attendances for 30 students.
So the mean number of attendances for each student is given by
$270 \div 30 = 9$ half-day attendances.

A4 Look at the attendance figures for 10M.
Does 9 give a fair idea of 10M's attendance record?

A5 What is the mean number of attendances for 10W?

A6 Which class had the better mean attendance figure?

Emma is still not satisfied. She decides to plot the attendances for each class on a number line.

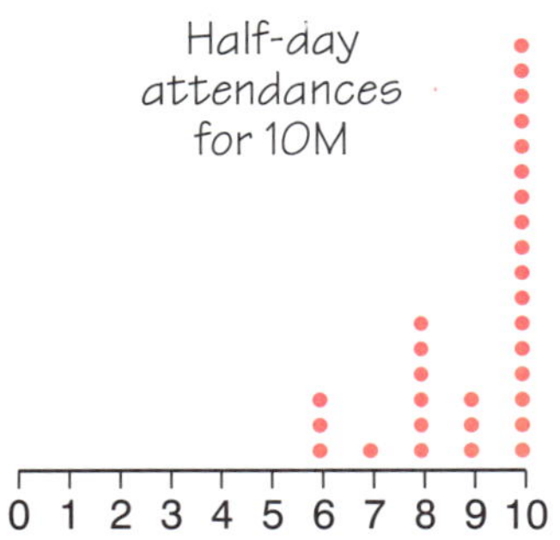

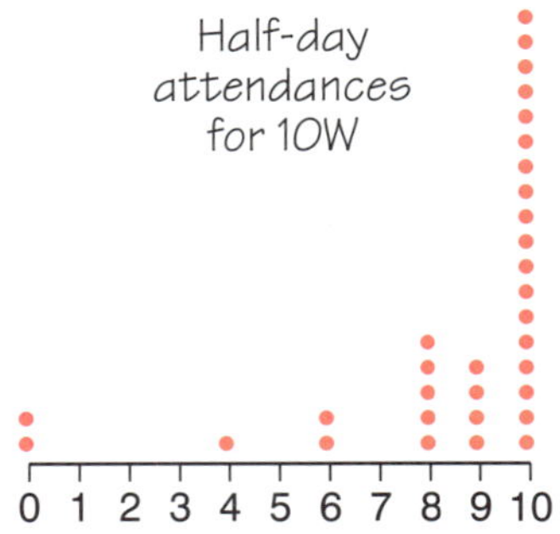

A7 From the number lines how many students attended 8 times in

(a) 10M (b) 10W?

A8 If you look at the attendance figures, there are far more even attendance numbers than odd. Why do you think this is?

For 10M the best attendance was 10 and the worst was 6.
The difference between these two figures is called the **range**.
So the range of 10M's attendances is 10 – 6 = 4 half-day attendances.
The range tells us how spread out the attendances are.

A9 What is the range of 10W's attendance figures?

A10 What do the two ranges tell you about the classes' attendance records?

A11 Here are the attendance figures for a week for two classes. Work together to make a short report giving your choice of the class with the better attendance record.
Give reasons for your answer.

10E, total number in class = 22
total number of attendances = 198

Frequency of attendances

Number of attendances	1	2	3	4	5	6	7	8	9	10
Number of students	0	1	1	1	0	0	0	0	1	18

10C, total number in class = 26
total number of attendances = 234

Frequency of attendances

Number of attendances	1	2	3	4	5	6	7	8	9	10
Number of students	0	0	0	1	1	0	1	3	6	14

B A better than average reaction

Before starting this section you each need to measure your reaction time five times. To do this you need worksheet G+ 4.
The reaction timer gives reaction times in hundredths of a second.

Don't forget to note down your reaction times!

This section is suitable for a group of three.

B1

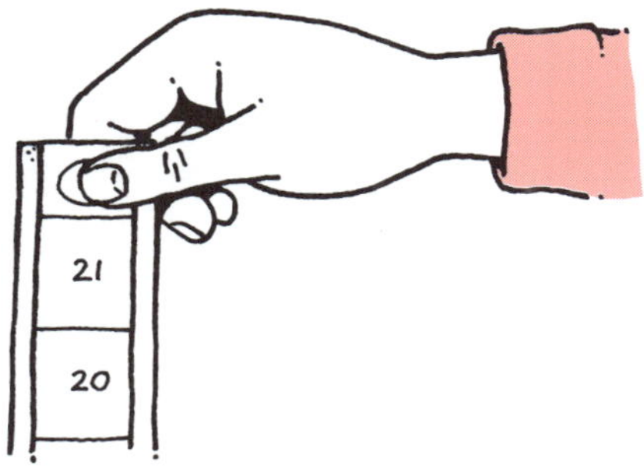

Work together to answer these questions.

(a) Who had the best single reaction time?

(b) How many hundredths of a second was it?

(c) Who had the worst single reaction time?

(d) Whose set of five reaction times was the best overall?

(e) How did you decide on your answer to (d)?

B2 What happens to your reaction times when you use your other hand to catch the ruler? Test each other again.

Work together on a short report saying what you did and noticed.

discussion point

How would you put these times in order?

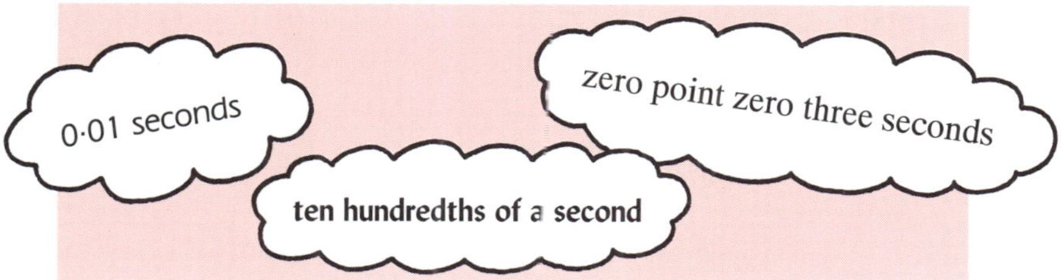

Sarah and some friends have just found their reaction times.
Here are their results.

All reaction times were measured in hundredths of a second.
We took turns at testing each other.

Here are the results.

Sarah – 17, 13, 12, 14, 16
Amal – 11, 10, 17, 18, 14
Jon – 25, 14, 12, 10, 10

When the reaction times were put in order, they looked like this.

Sarah – 12, 13, 14, 16, 17
Amal – 10, 11, 14, 17, 18
Jon – 10, 10, 12, 14, 25

discussion points

Look carefully at the results from Sarah's group.

- Who had the best set of reaction times?
- Does it help you decide if the results are put in order?

Sarah found this in one of her older sister's statistics books.

Putting results in order can sometimes help you to find an average or typical figure. For example, with results 14, 15, 20, 22, 40, the 40 might have been a fluke. So it might be better to use the middle result. This is 20. The middle result is called the **median**.

A measure of spread is given by the **range**. This is the difference between the biggest and smallest result. For example, for 14, 18, 19, 20, 24, the range is 24 – 14, which is 10.

Does using the median help you with the first 'discussion point' question?

B3 Find the median reaction time for Sarah, Amal and Jon.
Does the median help with the first 'discussion point' question?

B4 Find the median and range of your own reaction times collected in question B1. Now look back at your answers to B1(d) and (e). Do you want to change your minds?

Sarah has a problem with medians ...

For Sarah's times of 7, 8, 9, 10, 12, 14, the middle two numbers are 9 and 10. The number half-way between these two is $9\frac{1}{2}$ or 9·5, so this is the median.

B5 Find the median of these reaction times (all are measured in hundredths of a second).

(a) 10 20 15 5 5 (b) 15 5 15 20

(c) 20 15 10 10 (d) 10 20 8 12 16

Challenge

Make up a set of five reaction times (all measured in hundredths of a second) which have:

(a) a range of 4 and a median of 10

(b) a range of 6 and a median of 16

(c) a range of 9 and a median of 12

Which of the sets of results gives the best set of reaction times? Give reasons for your answer.

Estimating capacity

These activities are for one or two people.

The amount that a container holds is called its **capacity**.

The capacity of the bottle of water is 1 litre or 1000 millilitres.
The capacity of the shampoo bottle is 200 millilitres.

Millilitres are sometimes written as ml. You have probably seen ml on bottles at home.

1 Ask your teacher for about eight containers labelled A, B, C, …
Put them in order of capacity.

Now find the container labelled 1 litre.
Which of the other containers have capacities greater than one litre?
Which have a capacity less than one litre?

2 Estimate the capacities of thecontainers.
You could check your estimates using a measuring jug and water.

3 The capacities of sports bags, rucksacks and suitcases are often given in litres. Estimate the capacities of some people's bags in your class.
Don't check using water!

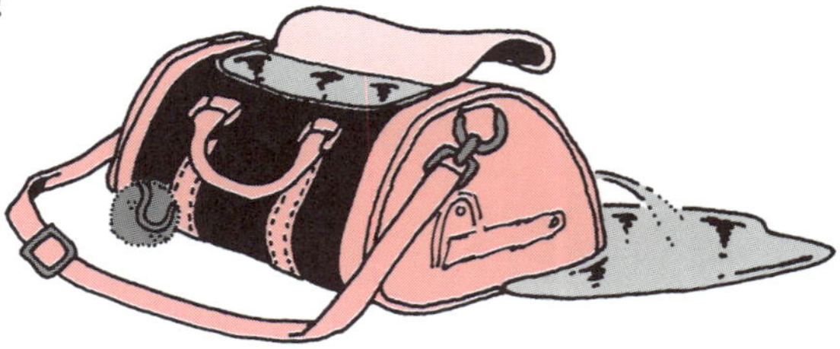

Challenge

How could you measure the capacity of a sports bag or rucksack?

4 What is the smallest capacity container (labelled with its capacity) you can find? What is the largest?

5 Shapes and sequences

A Matchsticks

Choy used some matchsticks to make these shapes.

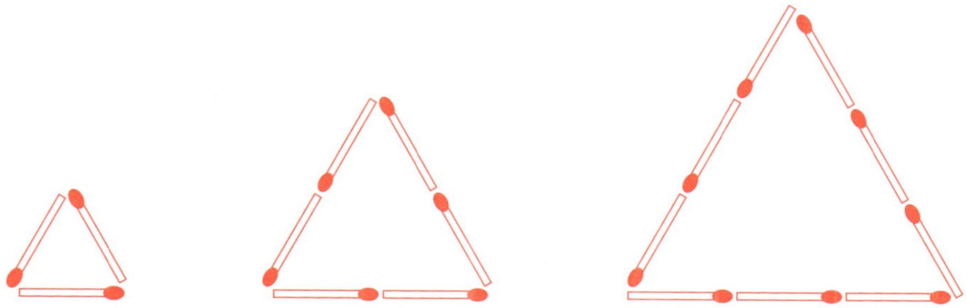

A1 How many matchsticks are used in Choy's first triangle?

A2 How many matchsticks are used in her second triangle?

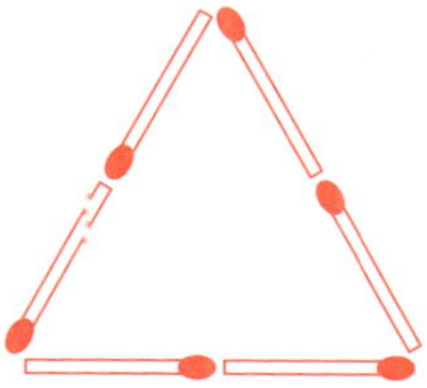

A3 How many matchsticks did she use in her third triangle?

A4 How many matchsticks will Choy need when she makes her fourth triangle?

Choy's shapes form a **sequence**.
The triangles get bigger every time and use more and more matchsticks.

A5 Make a list of the number of matchsticks Choy used in each triangle. (Include her fourth triangle as well.)

A6 What will the next three numbers in your list be? Check your answers by drawing the triangles.

A7 How many matchsticks will Choy need to make the tenth shape in her sequence?

A8 Describe what you notice about the numbers in the sequence.

Choy decides to make new shapes.
She uses the matchsticks to make squares this time.

Here are her first three squares.

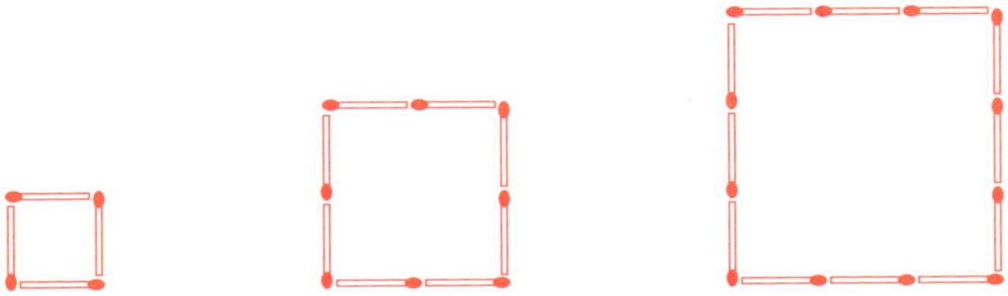

A9 How many matchsticks are used in the first square?

A10 Make a list of the number of matchsticks used in the first three squares.

A11 What will the next three numbers in the sequence be?

A12 The numbers in the sequence follow a rule.
Describe in your own words the rule for this sequence.

Challenge

Make some sequences of shapes from matchsticks.
(Draw the sequences of shapes if you cannot find any matchsticks.)

Try to find the rules for making your sequences.

B More sequences from shapes

Joanna watches Choy making the shapes with the matchsticks.
She starts making these solid shapes using multilink cubes.

B1 Make a list of the number of cubes used by Joanna in making the first five solid shapes in her sequence.

Check your answers by making the shapes yourselves.

B2 Describe the rule for Joanna's sequence.

B3 Mike decides to make some shapes from red tiles.

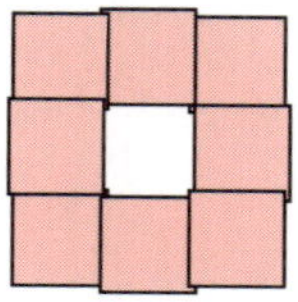

First shape

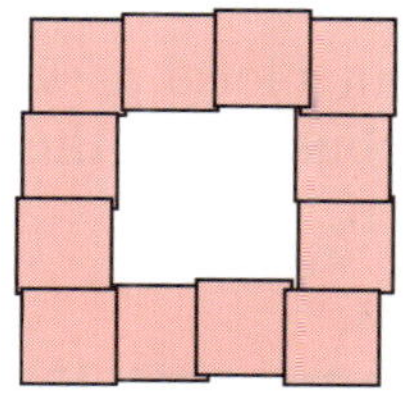

Second shape

Third shape

(a) How many tiles are there in each shape?

(b) What is the rule for the numbers in this sequence?

(c) How many tiles will there be in the sixth shape in this sequence?

B4 This is another sequence of tiles.

1 tile

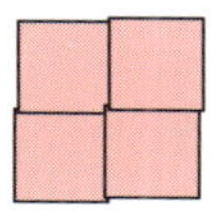

4 tiles

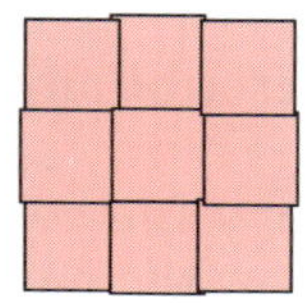

9 tiles

(a) How many tiles will there be in the next three shapes?
(If you are not sure, check your answers by drawing the next few shapes in the sequence.)

(b) What is the rule for the numbers in this sequence?

B5

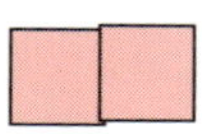

2 tiles

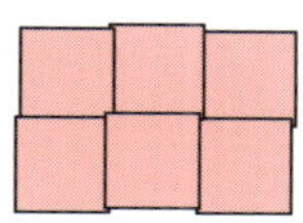

6 tiles

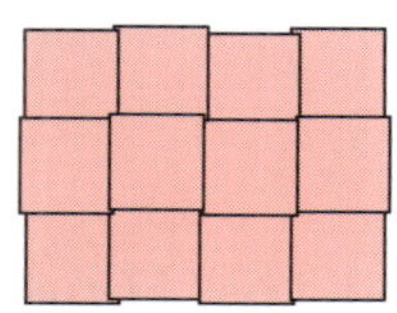

12 tiles

Look at the shapes carefully to see how they are built up.

(a) Describe how to make the fourth shape from tiles.

(b) Make a list of the number of tiles in the first four shapes.

(c) How many tiles will there be in the fifth shape?

(d) What is the rule for the numbers in this sequence?

B6 (a) Use multilink cubes or tiles or matchsticks to make some shapes which follow a simple rule.

(b) Describe the rule for your sequence.

(c) Show a friend your shapes. Do not describe your rule.
Ask your friend to make the next shape in the sequence.
Check that it fits your rule.

C Up the garden path

Sandra is making new rose beds in a local park.
Each new rose bed has one more bush than the row before.
Each rose bed has a path around it.

Here are some of Sandra's designs.

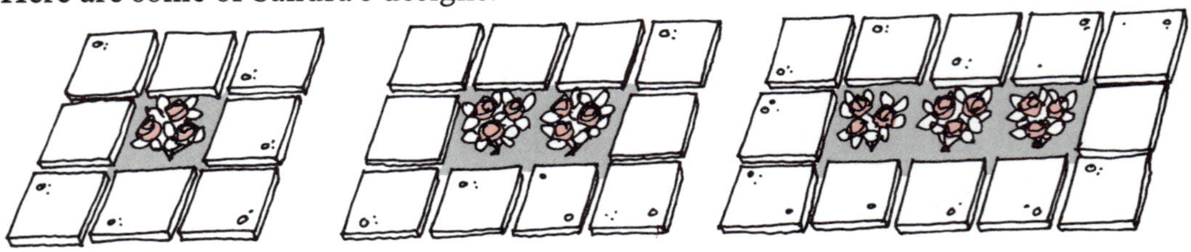

C1 Draw the fourth rose bed.

C2 Describe any differences between the third and fourth rose beds.

C3 Copy and complete this table to show the number of rose bushes and the number of paving slabs required for each rose bed.

Bushes	Paving slabs
1	
2	
3	
4	
5	
6	

C4 Sandra has 11 rose bushes to put into a rose bed.
How many paving slabs will she need for the path that goes around the rose bed?

When Sandra puts the path around the rose bed, she puts the path either side of the bushes first.

Then she finishes off with the ends.

Sandra can use a number machine chain to tell her how many paving slabs are needed for each rose bed.

C5 Which of these machine chains could Sandra use to help her?

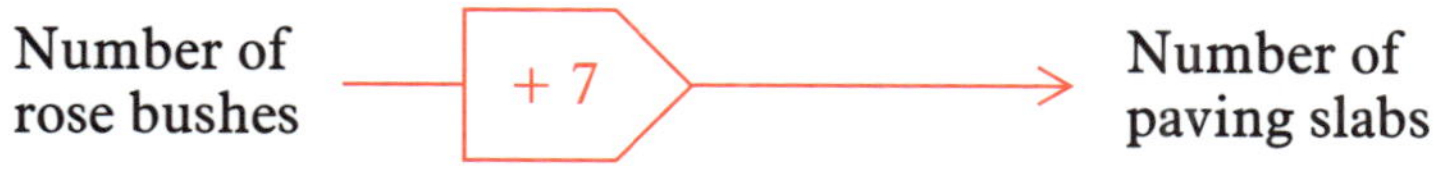

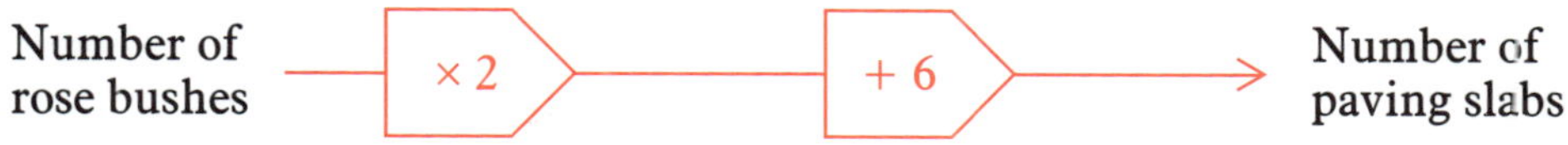

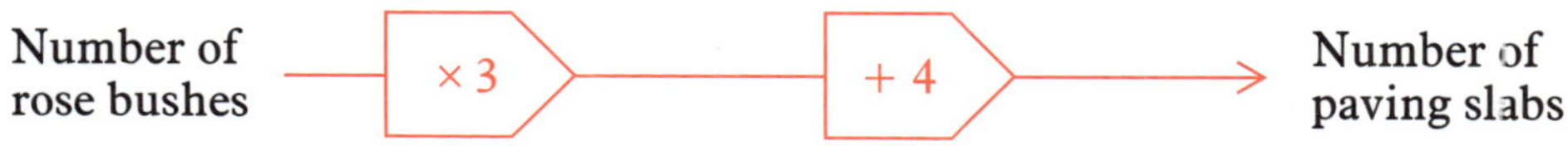

C6 Use the machine chain to check your answers for questions C3 and C4.

C7 How many paving slabs will Sandra need to make a rose bed for 100 rose bushes?

C8 Sandra makes a rose bed using 26 paving slabs. How many roses will she need?

Challenges

- Copy and complete this table for the number of matchsticks in Choy's triangle sequence in **A5**.

  Write a machine chain for the number of matchsticks used in each of Choy's triangles.

Number of matchsticks on each side	Number of matchsticks needed
1	
2	
3	
4	

- Choy produced a sequence of numbers when she made squares with matchsticks.

  Show the information in a table and then write a machine chain to help find the number of matchsticks needed to make the different squares.

Boxes

You need a pencil, squared paper and scissors.

This flat shape may be folded up to make a cube.

We say that the flat shape is a **net** of the cube.

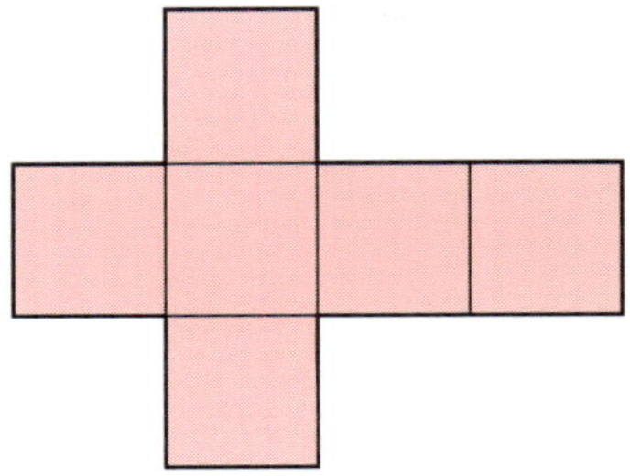

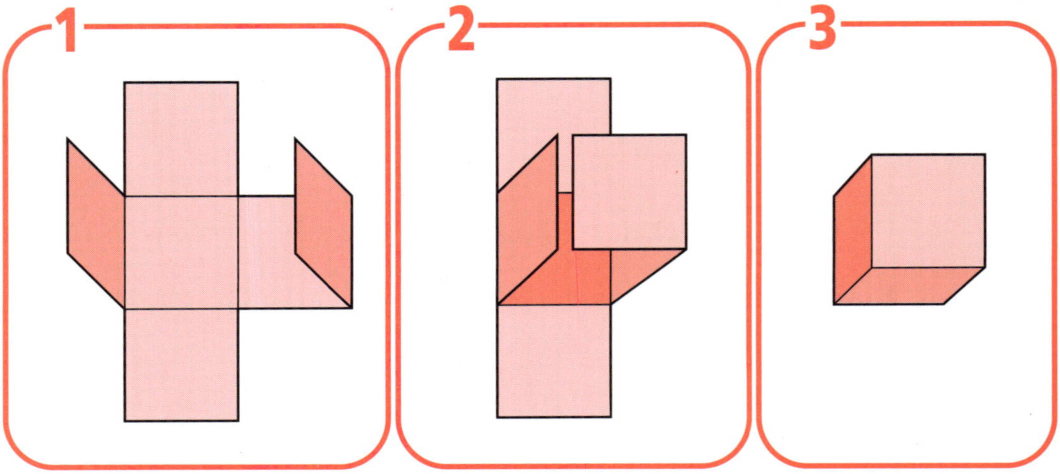

Here are some more nets of cubes. Imagine them being folded into cubes. If you are not sure, draw the nets on squared paper, cut them out and fold them into cubes yourself.

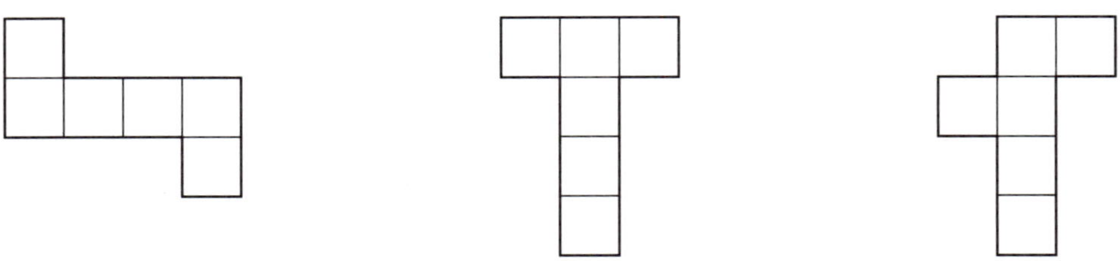

1 How can you tell – just by looking – that these cannot possibly be nets of cubes?

(a)

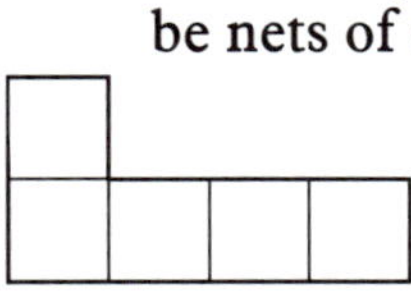

(b)

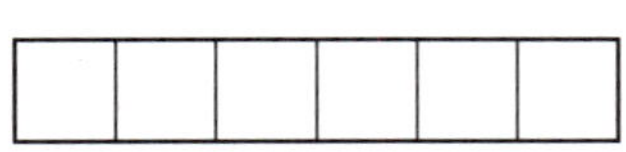

(c)

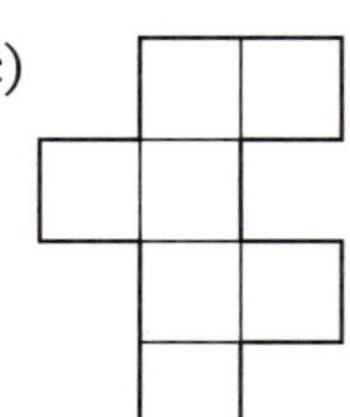

2 Which of these are nets of cubes and which are not?
(If you get stuck, copy the nets onto squared paper.)

(a) 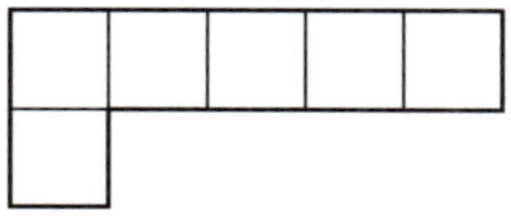

(b) 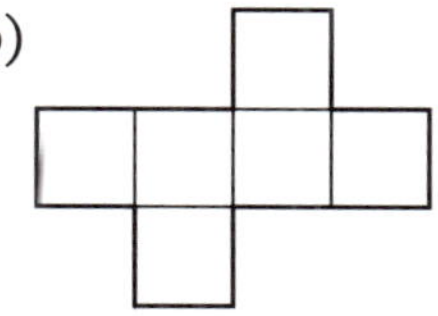

(c) 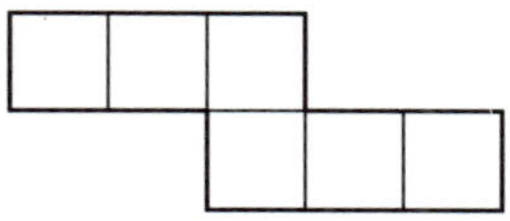

(d) 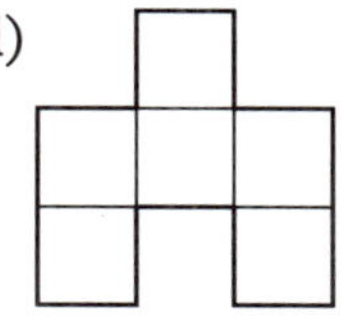

3 Is this true?

A line of four squares with a square on each long side always gives the net of a cube.

For example, is this a net for a cube? Experiment for yourselves.

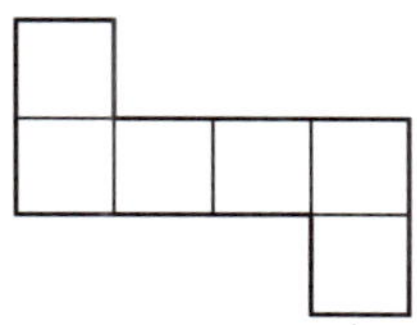

4 Is it possible to make a net for a cube from six squares when five of the squares are in a line?

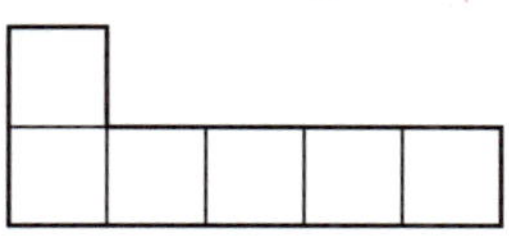

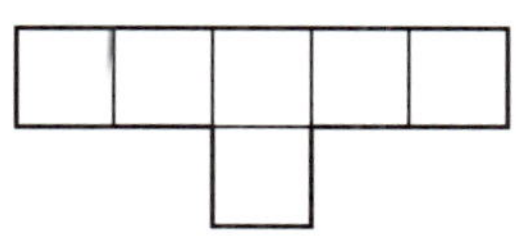

5 Draw as many shapes as you can by joining five equal squares edge to edge.

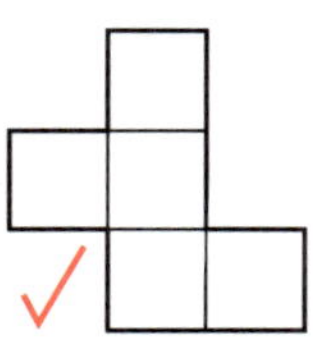

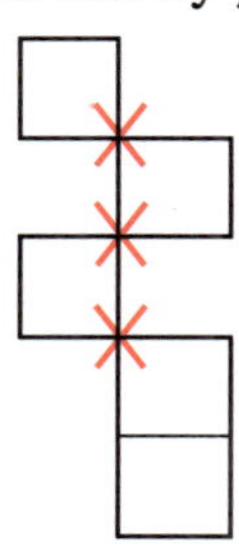

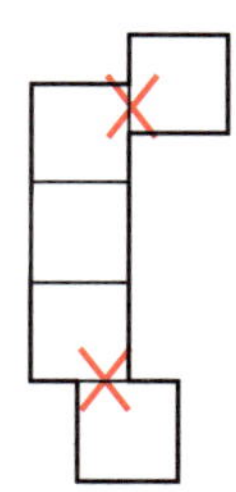

Decide if each shape can be folded to make a box without a lid.

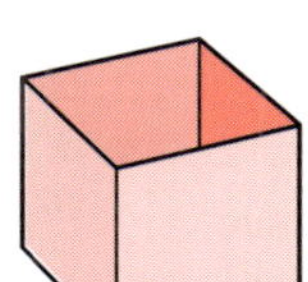

Compare your results with someone else's.

6 Use this flow diagram (or flow chart) to check your answers to **A5**.

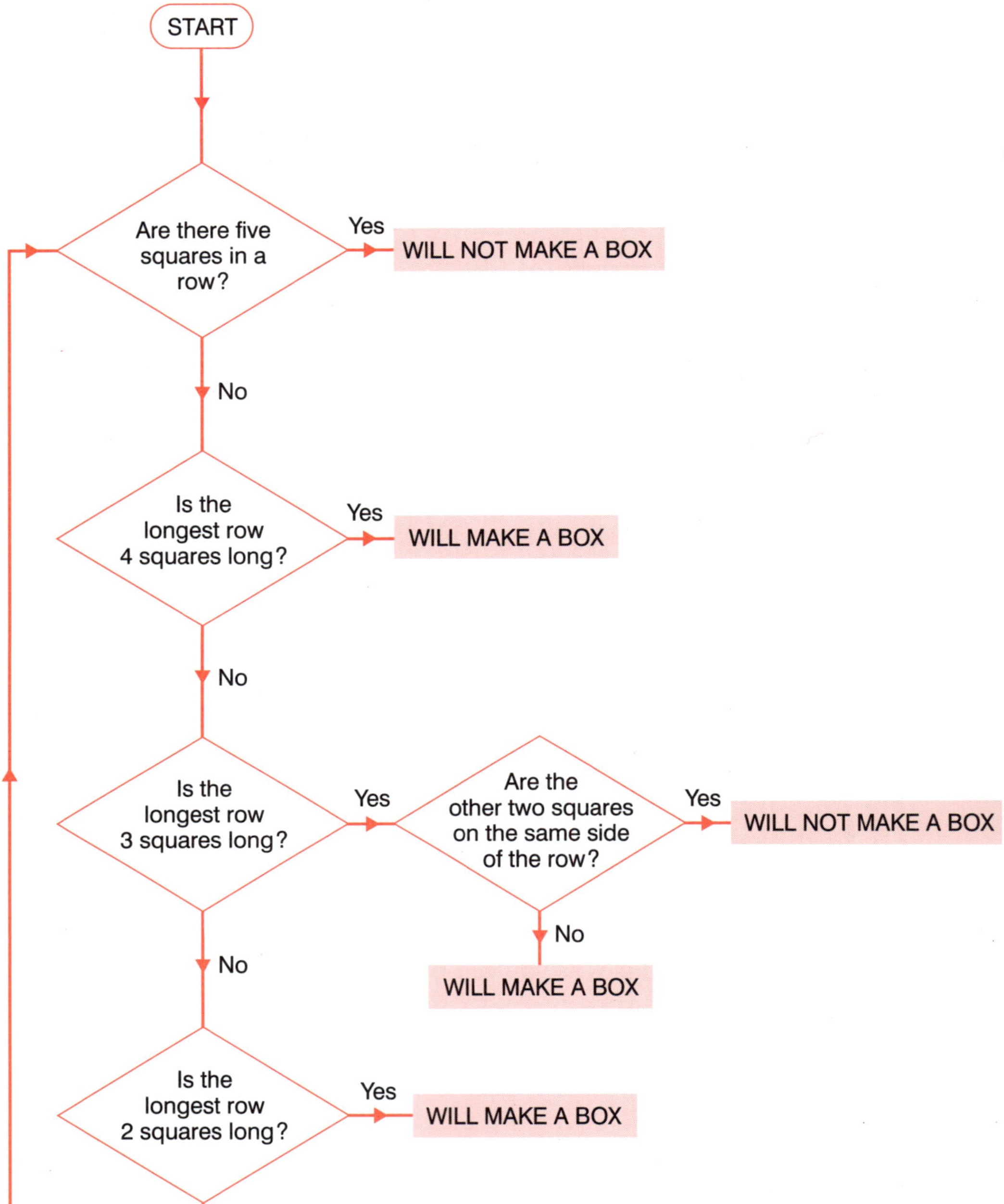

6 Leftovers

A Division

Egg boxes hold six eggs.
Suppose we want to find out how many boxes are needed to pack 25 eggs.

Suppose we want to find how many netball teams of seven players can be made from 20 people?

These are both division problems.

To find the number of egg boxes, we have to find how many sixes there are in 25.

The calulator shows the result of 25 ÷ 6.

We don't have a whole number answer, so to find the number of egg boxes we have to think about whether to round up or down.

It is no good having eggs left without a box, so we need five boxes.

We have to round up.

To find out the number of netball teams, we have to work out how many sevens there are in 20.

The calculator shows the result of 20 ÷ 7.

We cannot have a team of less than seven, so we can only make 2 teams.

We have to round down.

It does not matter that 4·16... is only a little above 4 or that 2·85... is very nearly 3.

A1 How many 20p stamps can you buy for £1?

A2 How many 19p stamps can you buy for £1?

A3 I have ten 19p stamps. I have to send £1 in stamps for a special offer book. How many stamps will I have to send?

A4 A costume for a play takes four metres of fabric. How many costumes can be made from 15 metres?

A5

Seven people are knitting squares for a blanket for charity. 500 squares are needed.

(a) How many squares will have to be knitted by each person to be sure to have enough?

(b) How many extra squares will have been knitted?

A6 Boxes 9 cm wide are to be stored on a shelf. The shelf is 80 cm long.

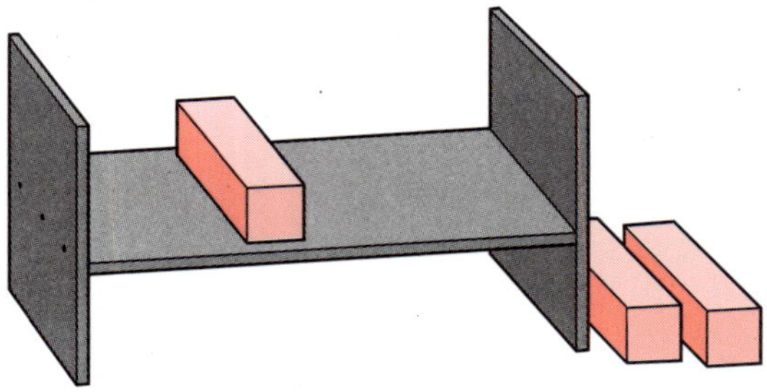

How many boxes fit on the shelf?

A7

A school trip is being arranged for 32 people.
They are going to travel by minibus.
Each minibus holds 12 people.
How many minibuses are needed?

A8 260 people go on a school trip. Each bus can carry 36 people.

(a) How many buses are needed?

(b) All but one of the buses are full.
How many full buses are there?

(c) How many people are there on the partly filled bus?

(d) How many more people can go without hiring another bus?

Challenge

Find some more real-life situations involving division where the answers need to be (a) rounded up or (b) rounded down.

Try to find two of each type.

B Remainders

Six friends are playing a card game.
They first have to share out 52 cards equally and put any left-over cards on the table.

How many cards do they each get?

How many cards are left over?

If you enter 52 ÷ 6 = on the calculator, you get a display of

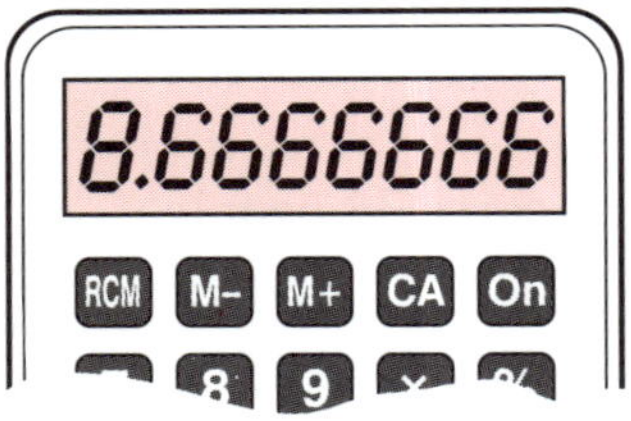

or

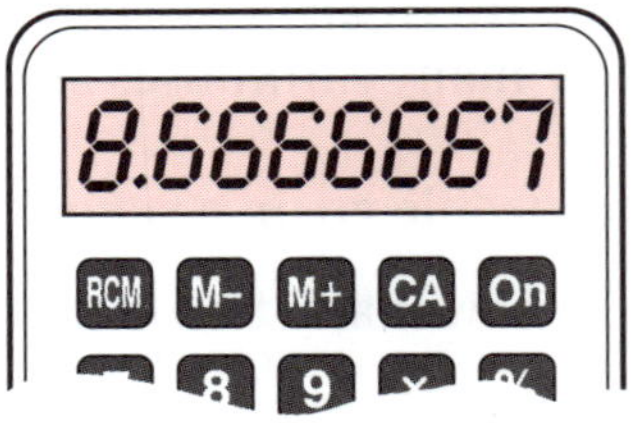

When you deal the cards out, each person gets 8 cards and there are some remaining (left over).

One way to find the remainder is to keep taking away 6 (starting with 52), until the answer is less than 6.

Experiment with your calculator to try to find an easy way to keep taking 6 from 52.

You could try [5] [2] [–] [–] [6] [=] [=] or

[5] [2] [–] [6] [=] [=]

Look at what is happening to the display.

B1 Work out how to use your calculator to count in fives starting from zero, and only pressing [5] once. See your teacher if you cannot do this.

How could you make it count backwards in ones from 10?

B2 Use your calculator method to find how many times you can take 13 away from 91. Check your answer by dividing.

B3 Work out the remainders for these calculations.

(a) $83 \div 17$ (b) $49 \div 13$

(c) $260 \div 37$ (d) $572 \div 46$

B4

There are 164 rooms to be cleaned in a hotel.
Josie, the senior cleaner, decides to share the rooms equally between the 14 assistant cleaners. She will clean the remainder herself.

(a) How many rooms will each assistant clean?

(b) How many rooms will Josie clean?

B5 Here are 8 classes and 225 pupils in Year 7.
The headteacher wants the pupils shared equally between the classes.

How many pupils will be in each class?

When one number divides exactly into another number (in other words there is no remainder) we say that the second number is **divisible** by the first.

So 24 is divisible by 3 and 8 because they both divide exactly into 24.

There are methods to find out if numbers are divisible by other numbers.

B6 By looking at a number how can you tell if it is divisible by five? (Hint. Ten will divide exactly into a number if it ends in a zero.)

B7 How can you tell if a number is divisible by two?

B8 For each part, find at least one number which

(a) is divisible by 5 and 7

(b) is divisible by 6 and gives a remainder of 8 when divided by 10

(c) gives a remainder of 1 when divided by 2 and a remainder of 2 when divided by 3

(d) gives a remainder of 3 when divided by 4 and a remainder of 3 when divided by 5

Make up and test some puzzles like these for yourselves.

The remainder game

This is a game for two, three or four players.

You will need to cut out the 30 number cards on worksheet G+ 5.

- Each player writes down any four numbers between 10 and 100.

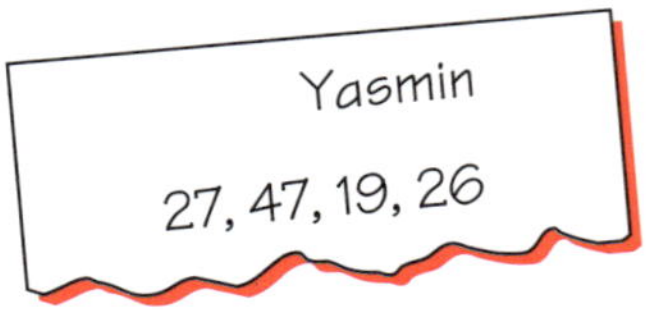

- Deal out six number cards to each player.

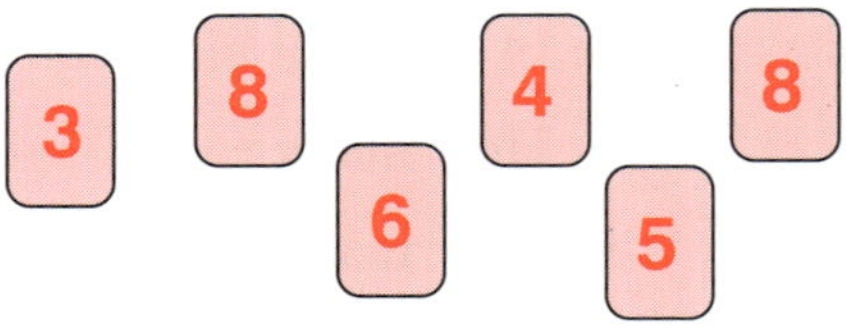

- The first player chooses one of the numbers from their list. They cross off this number and divide it by one of their number cards. Their score for the round is the remainder.

- Return the number card to the pack.

Yasmin chose 19 and divided by 8.
The remainder is 3.

- Keep a record of the score like this.

	number card	Yasmin score
19	8	3
26	4	2
27	6	3
47	3	2
Total		10

- Players take turns until they have used up all their four numbers. The highest total score wins.

Flying away from it all

This is for two or three to do together.

More and more people are flying.

This **pictogram** shows the number of people flying to the USA from Britain each year from 1980 to 1987.

The figures have been rounded to the nearest million.
Each symbol ✈ stands for 1 million passengers.

1980	✈ ✈ ✈ ✈
1981	✈ ✈ ✈ ✈ ✈ ✈
1982	✈ ✈ ✈ ✈ ✈ ✈
1983	✈ ✈ ✈ ✈ ✈
1984	✈ ✈ ✈ ✈ ✈ ✈ ✈
1985	✈ ✈ ✈ ✈ ✈ ✈ ✈
1986	✈ ✈ ✈ ✈ ✈ ✈ ✈ ✈ ✈
1987	✈ ✈ ✈ ✈ ✈ ✈ ✈ ✈

1 Use the pictogram to answer these questions.

(a) How many million passengers flew to the USA from Britain in 1983?

(b) In which year did most people fly from Britain to the USA?

(c) How many passengers was this?

2 Here are the numbers of passengers flying from Britain to France each year from 1980 to 1987, to the nearest million passengers.

Year	1980	1981	1982	1983	1984	1985	1986	1987
Millions of passengers	3	3	3	3	4	4	4	4

Draw a pictogram to illustrate these figures. Decide for yourselves what symbol to use for 1 million passengers.

3 Stuart says that the two pictograms show that more people have their holidays in the USA than France.

Do the figures really show this?
Give some reasons for your answers.

You need

- worksheets G+ 6, G+ 7, G+ 8 and G+ 9
- scissors
- a watch measuring in seconds
- a computer loaded with a data handling program
- the data file 'Florida'
- a brochure about holidays in Florida

Worksheets G+ 6 and G+ 7 show the details of some holidays in Florida.

Each holiday is based on a certain hotel. For example, the Delta Court of Flags is a four-star hotel. It costs £674 per person for two weeks.

The hotel has a swimming pool and a restaurant which serves free children's meals. It also has free car hire.

DELTA COURT OF FLAGS

****	£674
swimming pool	✓
restaurant	✓
free children's meals	✓
free car hire	✓

Cut out the cards on worksheets G+ 6 and G+ 7. Shuffle them.

Worksheet G+ 9 shows the same information in a table.

Name	Star rating	Cost per person (£)	Swimming pool	Restaurant	Free children's meals	Free car hire
Delta Court of Flags	4	674	yes	yes	yes	yes

Time how many seconds it takes each of you to answer each of these questions.
One of you use the cards, the other the table.

4 Which is the most expensive holiday?

5 How much does the cheapest holiday cost?

6 Which of the hotels have four stars?

7 Which is the most expensive three-star hotel?

8 Which is the cheapest two-star hotel?

9 How many four-star hotels offer free car hire?

discussion point

Which was quicker (or easier) to use, the cards or the table?
Why?

10 Use your own brochure to find details of some more Florida holidays. Add these to your table and fill in some more cards.

Time each other answering questions like the ones on the last page. You may choose cards or table. Make a game of it!

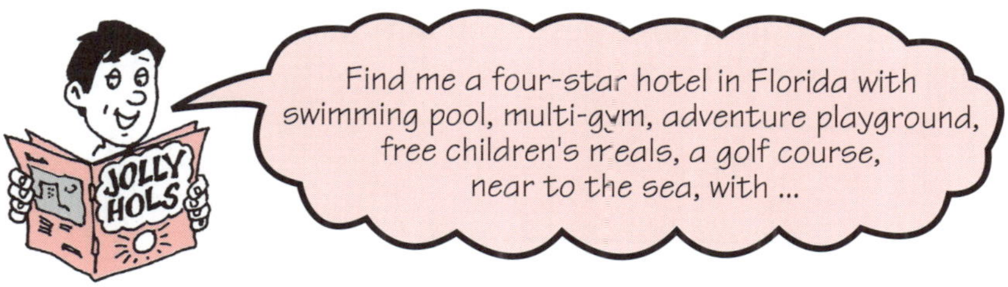

There were probably only about 25 holidays on the cards and in the table.
Imagine having a look through a thousand cards!
This is where a computer comes in useful.

Ask your teacher to load a data handling program together with the data file called 'Florida'.

You could try to figure out how to use it by experimenting for yourselves, or ask your teacher or a friend for help.

Use the program to check your answers to questions 1 to 9.

Ask your teacher to show you how to enter your information on hotels into the data file.

When you have entered all your information (or **data**), check that it's OK. The best way to do this is to look it up on the computer!

Make up a game about finding information like the one you played before. This time use the computer.

7 The results prove it!

This is a chapter for two or more people.

A The play street

Eastbourne Street is in a large town.

There are no parks nearby so children play in the street.
Over the last few years several children have been hurt in road accidents.
The people on the street would like it to be blocked to traffic at one end.

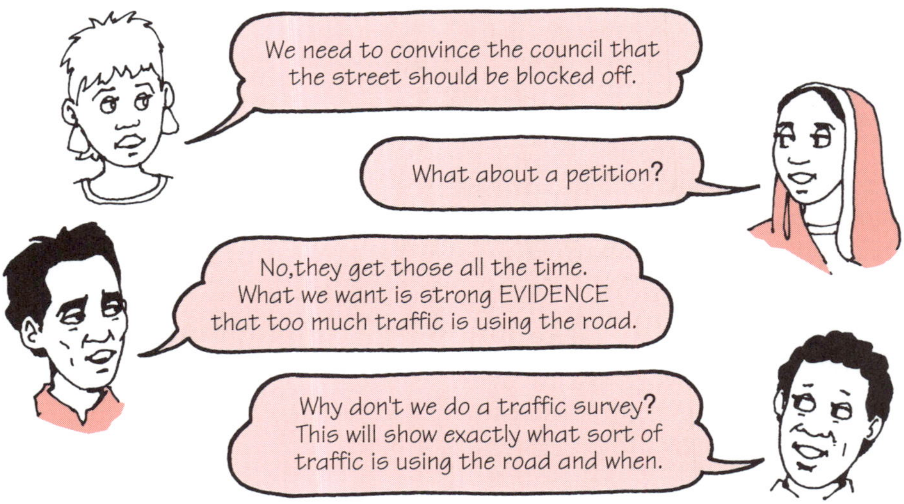

The next week Claire did a **survey** of the traffic going along the street.
Here is her **data collection sheet.**

Vehicles using Eastbourne Street on a Monday

Type	Tally	Total
Cars	卌 卌 III	13
Lorries	卌 III	8
Others	卌 卌 卌 III	18
		39

A1 Claire's is a good first attempt (or **pilot survey**).
(A pilot survey is a trial survey which lets you know what sort of numbers or data to expect.)

There are at least four ways in which Claire's survey sheet could be improved. Discuss what these might be. To help you, here are some comments made by her friends.

Claire's friends agreed to help her collect the data.
She decided to make a new data collection sheet.

She paid attention to what her friends had said.
Claire typed this new data collection sheet.

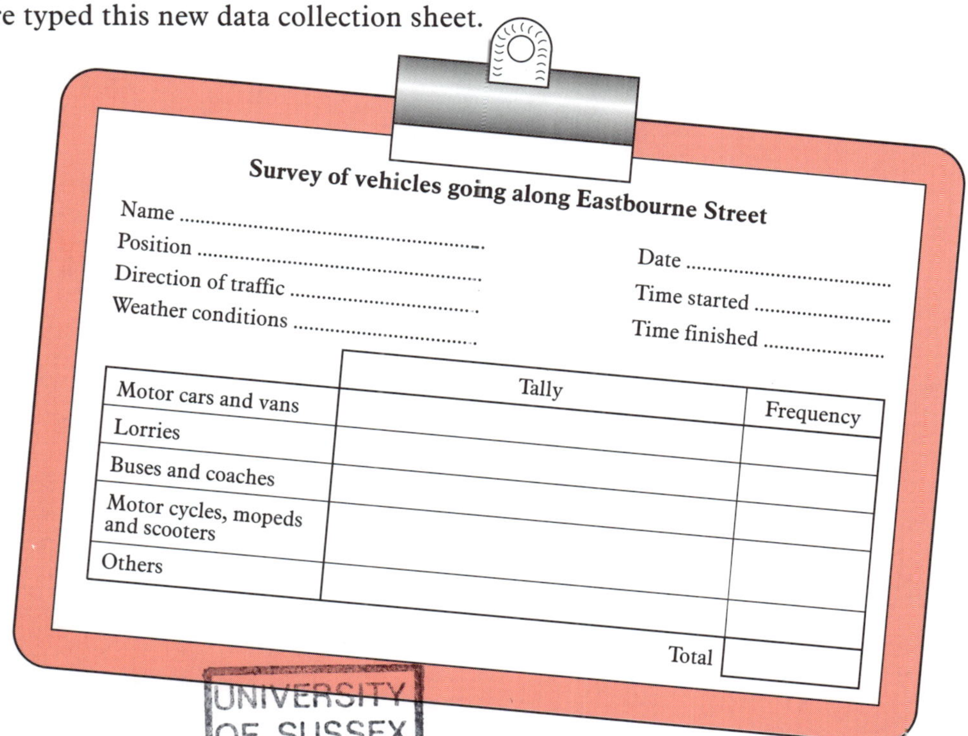

Survey of vehicles going along Eastbourne Street

Name ....
Position ....
Direction of traffic ....
Weather conditions ....

Date ....
Time started ....
Time finished ....

	Tally	Frequency
Motor cars and vans		
Lorries		
Buses and coaches		
Motor cycles, mopeds and scooters		
Others		
	Total	

Claire and her friends were almost ready to begin collecting data. Before they started they needed to think about these points.

- Should they do the survey each day for a week or for a month?
- Which day, or time of the day, should they choose?
- How long should the survey last?

Discuss these points yourselves.
Can you think of anything else they need to consider?
In the end the group decided to do their tally from 2 o'clock to 3 o'clock each day for a week.
Here are all the survey results added together.

Traffic totals for Eastbourne Street
(1 hour a day for 7 days)

Cars	134
Lorries	77
Buses and coaches	15
Motor cycles	33
Vans	15
Others	6
1 horse box	
3 cranes	
2 road sweepers	

Total = 280

A2 About how many vehicles use Eastbourne Street between 2 p.m. and 3 p.m. each day?

A3 About how many lorries use Eastbourne Street between 2 p.m. and 3 p.m. each day?

A4 Sonja thought that it might have been better to do the tally from 9 o'clock in the morning till 3 o'clock in the afternoon. Do you think this is a good idea? Give reasons for your answer and discuss them.

A5 When would you have done the survey? Give one or two reasons for your answer.

Arun asked his sister Mira if she could display the results from the traffic survey. She drew a bar chart and a pie chart.

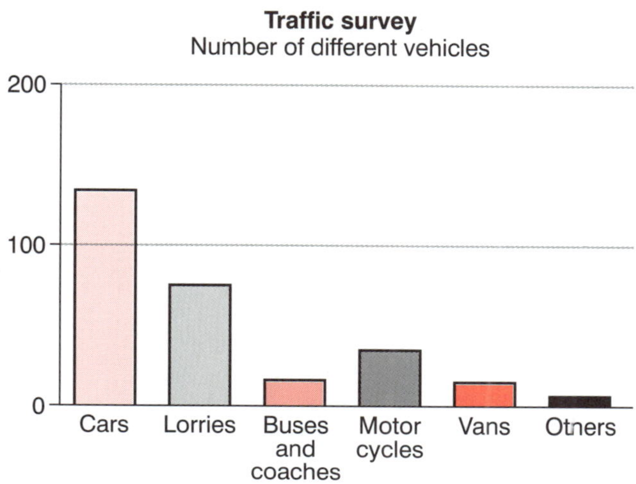

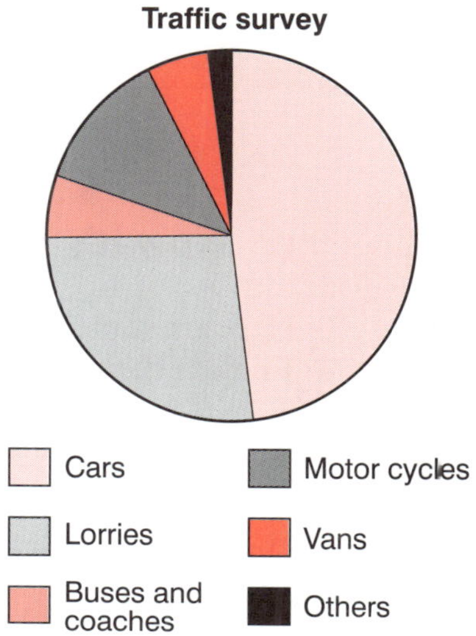

A6 Arun wanted to show that over half the traffic was cars, vans or lorries and that quite a large fraction of all the traffic was lorries. Which chart shows this more clearly?
Discuss this together.

A7 Mira and Arun showed the results of the traffic survey to the council. Underneath the results they wrote a **conclusion**.

> These results were obtained by doing a tally between 2 and 3 o'clock each day for a week.
>
> As you can see from the figures and chart Eastbourne Street has far too much traffic going along it. There are over 50 vehicles an hour passing along it. A lot of these are lorries.
>
> This amount of traffic is too dangerous – a lot of children live in the street.

Can you write a better conclusion for yourselves?
You can only use Claire's data on page 52.

A8 Try to think of any other data which Claire and her friends could have used to convince the council.

B Now it's your turn!

Here are some suggestions for surveys.
Look at them all and choose one.

At lunch time, how much traffic goes past the school gate?
Is it about the same each day?

Do young people prefer brown bread to white bread?

Do most cars only have one person in them – the driver?

Do girls spend more time on their homework than boys?

Which accidents happen most often in the home?

Which are the most popular teenage magazines?

Where would people most like to go for their dream holiday?

Do most people have two or more televisions in their house?

Before starting you need to talk through these questions.

- What are you trying to find out?
- What questions do you want to ask?
- Do you need to do a pilot survey?
- What information do you need on the survey sheet?
- Where and when are you doing the survey?
- How will you display your results?

Make a short plan and show it to your teacher.

Angles and shapes

These activities are for two people to do together.

Angles

You need an angle measurer.

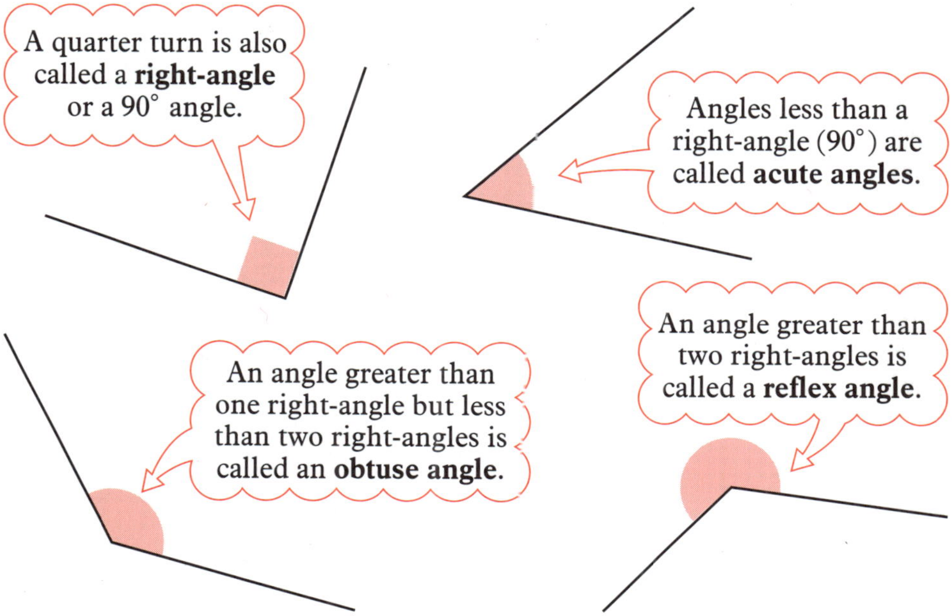

1 Estimate the size of the angles marked with a letter.
Say whether each one is an acute, reflex or obtuse angle.

Check your answers by measuring the angles with an angle measurer.
(The first one has been done for you.)

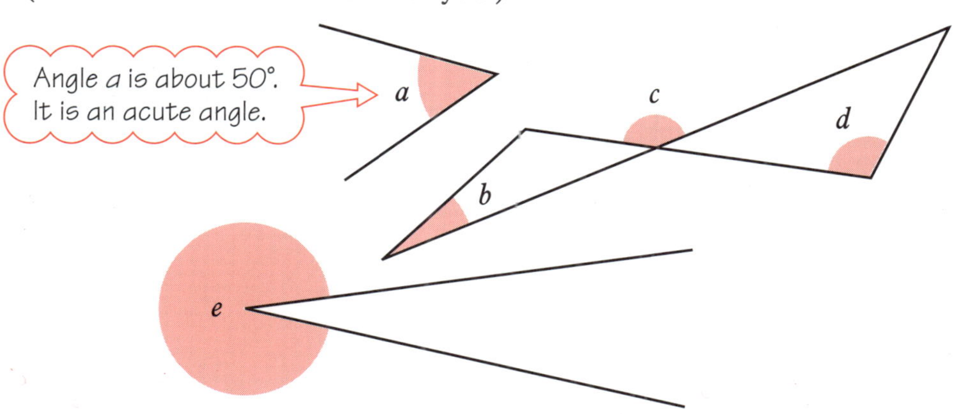

2 Copy and complete this table for the shapes labelled a, b, c and d.
The first one has been done for you.

Shape	Name	Number of acute angles	Number of obtuse angles	Number of reflex angles
a	triangle	2	1	0
b				
c				
d				

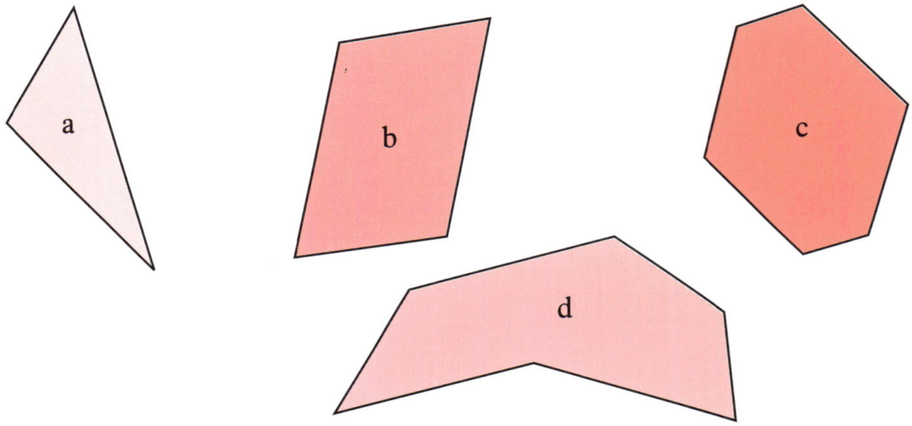

3 Draw one each of these.

- a triangle with three acute angles
- a quadrilateral with two acute and two obtuse angles
- a quadrilateral with three obtuse and one acute angle
- a five-sided shape (pentagon) with one reflex, two acute and two obtuse angles

4 What is the greatest number of acute angles a hexagon can have?
Investigate for yourselves.

What about an octagon or other shapes?

Shapes

You need worksheets G+ 10, G+ 11 and G+ 12, and scissors.

1 Here are some shapes.
Match the shapes with the statements in the bubbles.

You may not have met some of the shapes before.
It may help you to measure some of them.

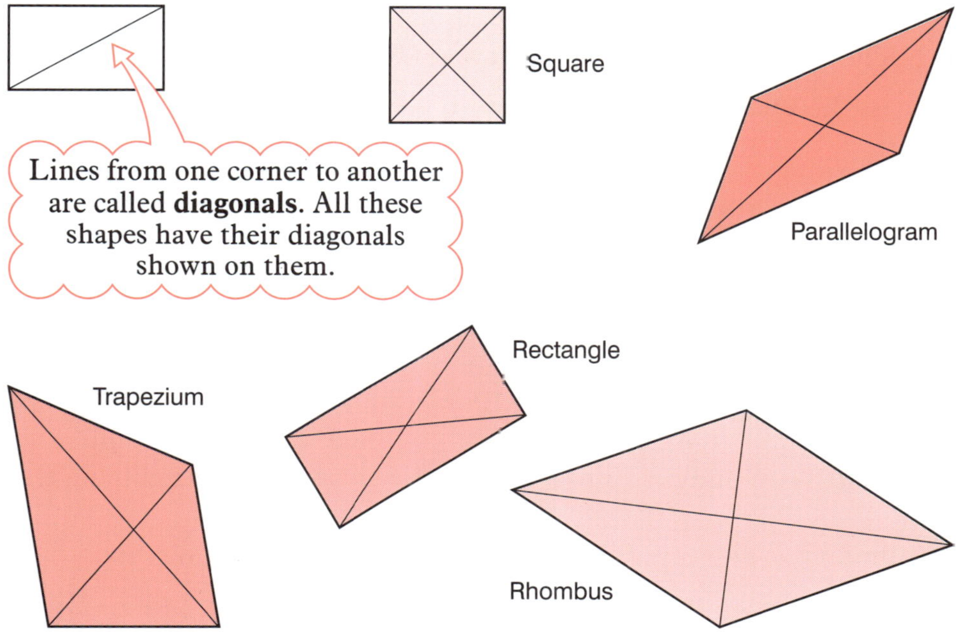

All sides are equal and all angles are equal; both pairs of opposite sides are parallel.

Only one pair of sides is parallel.

Opposite sides are parallel and equal; diagonals are equal in length.

Opposite sides are parallel; all sides, but not all angles are equal; diagonals cross each other at right-angles.

Opposite sides are parallel and equal; diagonals do not cross at right-angles and are unequal in length.

2

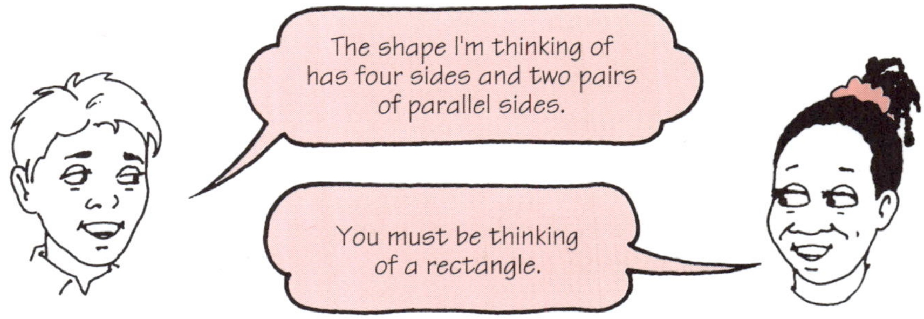

Is Nadia correct? What would you have said?

Discuss this question together.

3 The last two questions only looked at **quadrilaterals** or 4-sided shapes. There are also special sorts of triangle.

Here are four extracts from a maths dictionary.

equilateral triangle A triangle with all sides (and angles) equal.

isosceles triangle A triangle with two sides (and angles) equal.

right-angled triangle A triangle which has one angle a right-angle.

scalene triangle A triangle with no sides (or angles) equal.

Use the maths dictionary definitions to help you say what sort of triangle each of these triangles is.

(a)

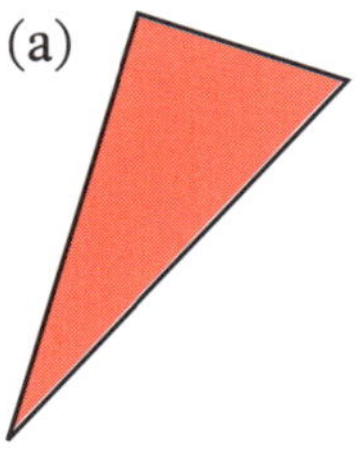

(b)

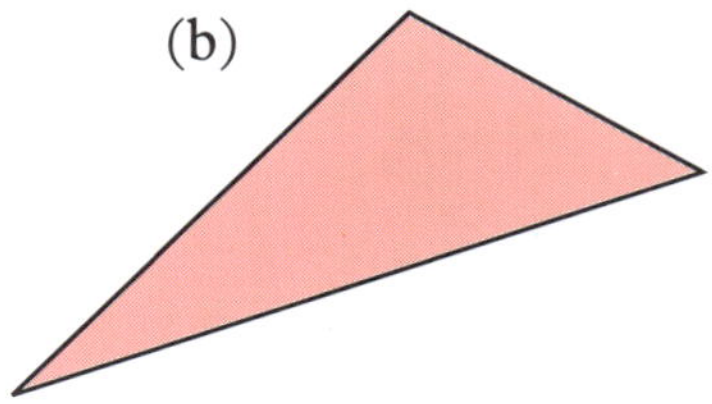

(c)

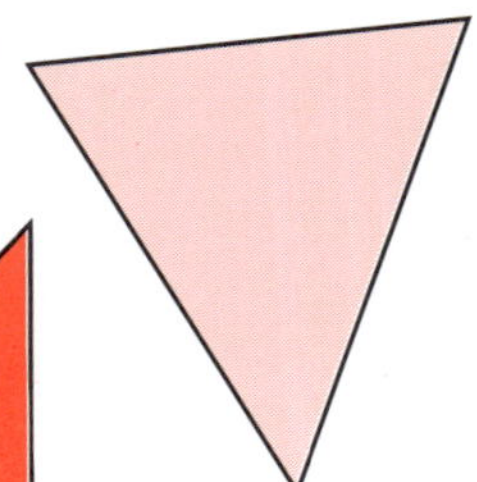

(f)

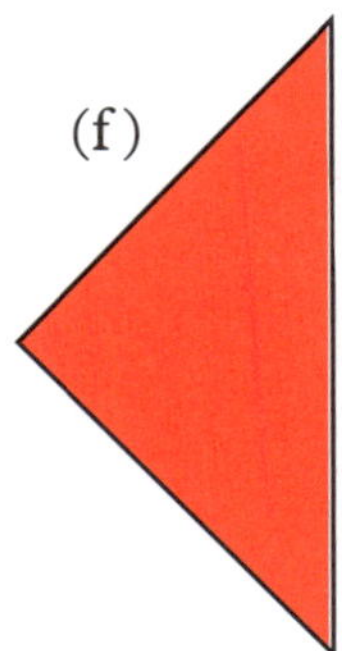

(e)

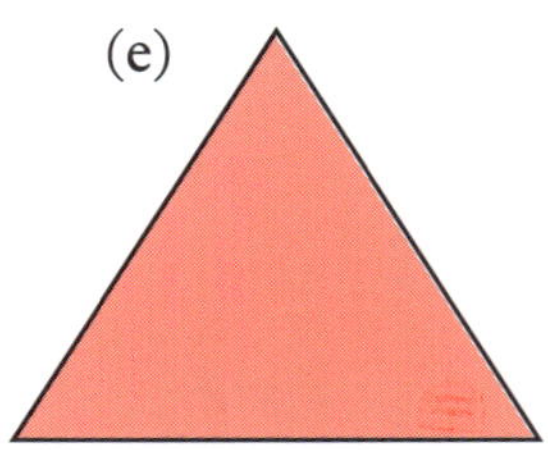

(d)

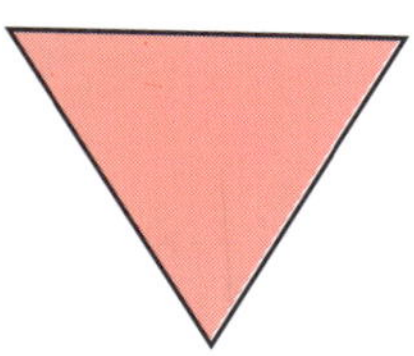

4, 5 and **6** These are on worksheets G+ 10, G+ 11 and G+ 12.

8 Negative numbers

A Number lines

Jumping

This is a game for two players.

You need worksheet G+ 13 for each player and a dice with the faces labelled + 1, + 2, – 3, – 1, – 2, – 3.

- Each player starts at the ZERO on their worksheet.

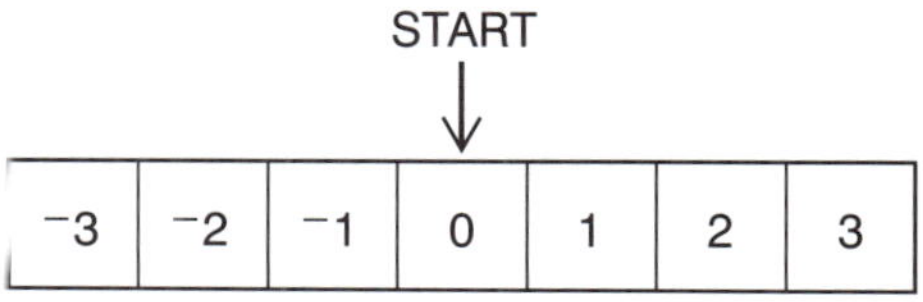

- Take it in turns to throw the dice.
  If you throw + 1, you move forward 1 place.
  If you throw – 3, you move back 3 places.
  Record the move on the worksheet.

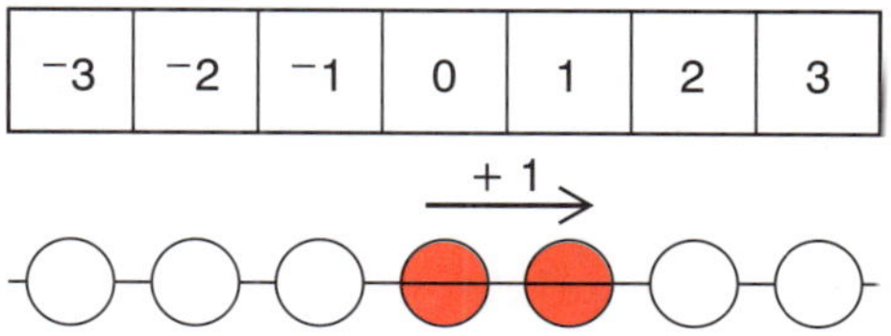

- On the second move, start at where you finished on the previous move.
  So if you throw + 1, then – 3 you move like this.

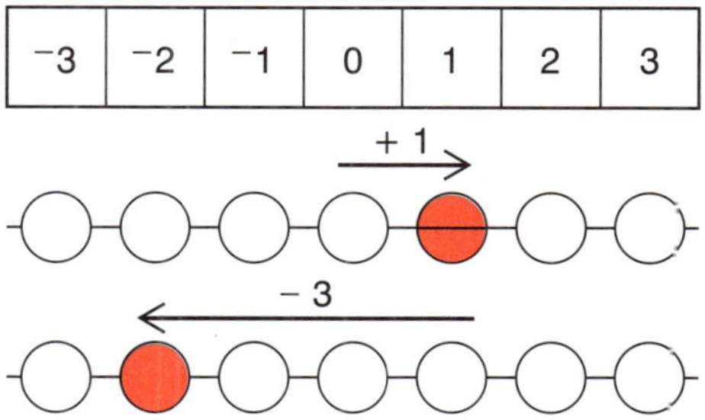

- Continue taking it in turns to throw the dice.
  Each player throws the dice 10 times.
- The winner is the player who ends up furthest from 0.
  Play the game three times.

A1 These are the first five throws for some players.
Work out where each player is after these five throws.

(a) Andy	+ 1	+ 1	− 3	− 1	+ 1
(b) Javed	+ 3	− 2	+ 2	− 1	+ 2
(c) Winston	− 1	− 2	− 1	+ 2	− 2
(d) Rachel	− 3	− 3	− 1	− 2	+ 1
(e) Shena	+ 1	− 2	+ 1	+ 1	− 3

A2 After their sixth throw, both Shena and Javed were on the same circle. Can you work out what they each threw?

A3 Misha has been playing 'Jumping'. He started on zero, threw the dice three times and ended back on zero.
List all his possible dice throws for this to happen.

You can use a number line to help you work out sums like $^{-}3 + 5$.

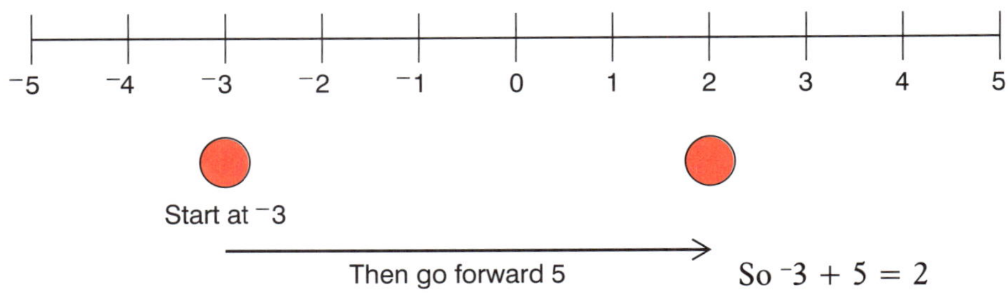

So $^{-}3 + 5 = 2$

A4 Copy and complete these:

(a) $^{-}2 + 3 = ?$ (b) $^{-}1 + 4 = ?$ (c) $^{-}5 + 8 = ?$

(d) $^{-}2 - 1 = ?$ (e) $^{-}3 - 1 = ?$ (f) $^{-}2 - 1 + 5 = ?$

A5 Try to work this out.

$^{-}18 + 41 - 28 = ?$

A6 Solve this number puzzle.

I start with 10.
I add 30,
subtract 50,
subtract 40,
and add 70.

What number have I now?

A7

A shop buys and sells plastic gnomes.
Here are the figures for the last month.

Week	Buy	Sell
one	5	6
two	0	2
three	7	8
four	6	5

(a) How many gnomes did the shop buy in the month?

(b) How many gnomes did the shop sell in the month?

(c) If the shop started the month with 10 gnomes, how many did it have at the end?

A8 Here are some number patterns.
Write down the next two terms in each pattern.
For each one explain how you worked out what they were.

(a) 7 4 1 $^{-}2$ $^{-}5$ …

(b) 30 20 10 0 $^{-}10$ …

(c) $^{-}10$ $^{-}8$ $^{-}6$ $^{-}4$ $^{-}2$ …

A9 Which of these is the correct way to say $^{-}2 - 5$?

B Temperature

B1

The temperature in Leeds was 25 °C but the temperature in London was only 18 °C.
How much cooler was London than Leeds?

B2 The temperature at 8 a.m. outside my house was 7 °C.
Now, at 11 a.m., the temperature has risen to 18 °C.
How many degrees has the temperature risen?

B3

If the temperature was 6 degrees Celsius below freezing, what would the reading on the thermometer show?

B4 The temperature in Anchorage is $^{-}13$ °C and it is 14 °C in Bristol, how much warmer is Bristol than Anchorage?

B5 The temperature outside my house at 6 p.m. was 2 °C.
By midnight the temperature had dropped 7 degrees.
What was the temperature at midnight?

B6 The temperature in a large block of ice is 11 degrees warmer on the outside than in the middle of the block.
The outside temperature is $^{-}3$ °C.
What is the temperature in the middle of the block?

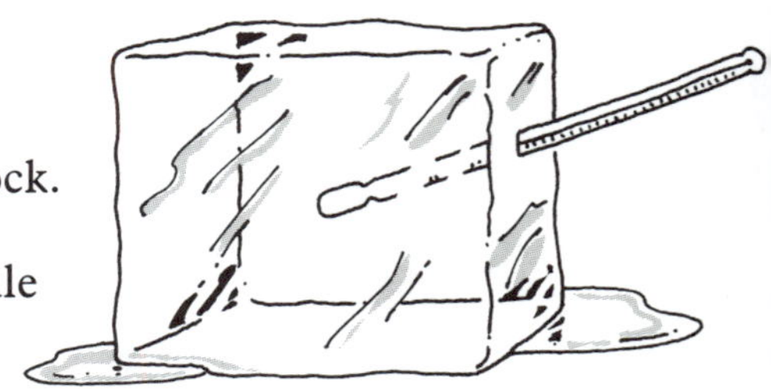

B7 The greatest change in temperature in Britain in a single day happened in Tayside in 1978. The temperature ranged from $^{-}7$ °C to 22 °C. By how many degrees did the temperature change?

B8 Low temperatures slow down the growth of germs.
The temperature inside a fridge should be between 0 °C and 5 °C.
In a freezer it should be no higher than −18 °C.

(a) Write down the temperatures showing on these fridge-freezer thermometers. Say whether each one probably comes from the fridge or freezer part of the fridge-freezer.

(i)

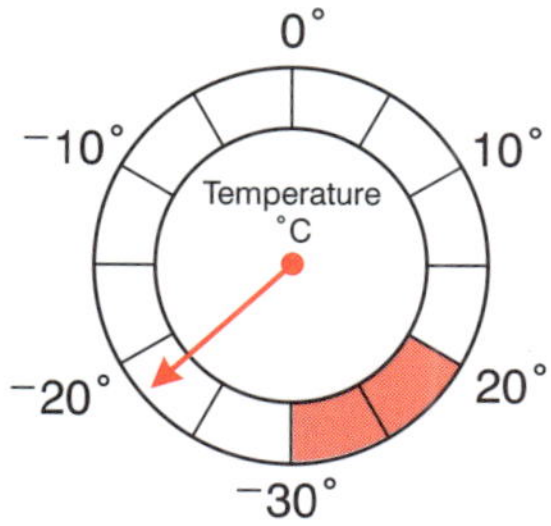

(ii)

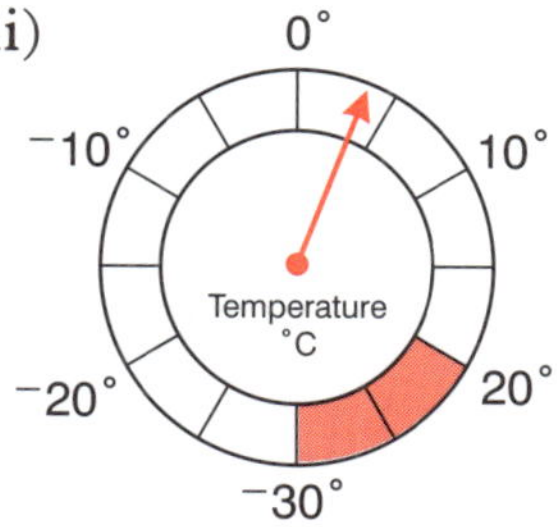

(iii)

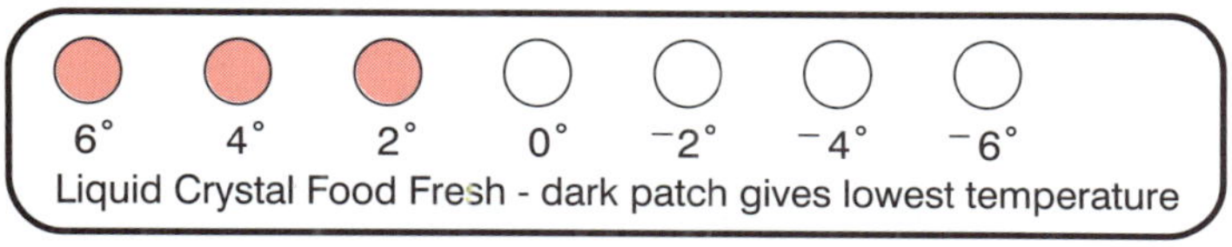

(b) A consumer magazine uses the number of kilograms of fresh food which can be frozen from room temperature to −18 °C in 24 hours to measure how quick a freezer is. About what temperature drop is this?

(c) If the electricity fails, food in the freezer may go bad. The magazine measures how good the freezer is by timing how long the temperature takes to rise from −18 °C to −9 °C when the power is turned off. What rise in temperature is this?

B9 The greatest temperature ranges are in Siberia. Temperatures as high as 37 °C and as low as −68 °C have been recorded. What range of temperature is this?

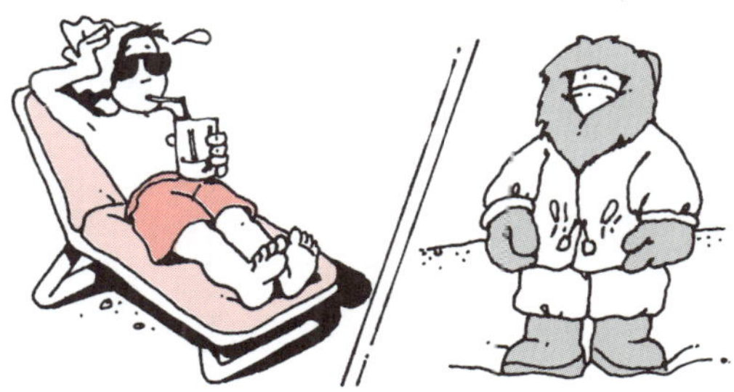

discussion points

- As dry air moves upward it cools by about 1 degree for every 100 metres rise in height. If ground temperature is 10 °C how high would you have to climb before the air was at freezing point (0 °C)?
- Which temperature is the lowest?

  −0·2 °C, 0·1 °C, $\frac{1}{100}$ °C, −0·8 °C.

Find out for yourselves

Try to find out as many of these as you can. Share your answers.

- The temperature inside a fridge – at school or at home
  The temperature inside a freezer – at school or at home
  Which is colder, the freezer or the ice compartment of the fridge?
  How much colder was it?

- All freezers (or ice compartments of fridges) have some stars on them.

  Find out what these stars mean.

- The temperature at which nitrogen becomes liquid
  Absolute zero temperature
  The temperature at which iron melts

- The surface temperature of Venus
  The temperature at the core of the Sun
  The lowest temperature recorded on the surface of the Earth
  The temperature in deep space
  The hottest and coldest temperatures on Mars
  The temperature at 'midday' and 'midnight' on the Moon

- The average summer and winter temperature in capital cities of the world. Which capital has the greatest temperature range?

- Can you find any other places where negative numbers are used, apart from temperatures?

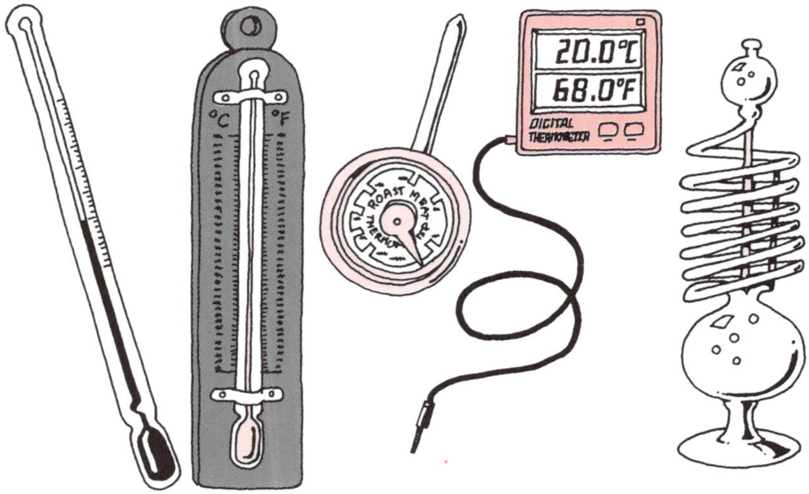

C Negative numbers on a calculator

C1 Use your calculator to work out these:

(a) $4 - 6 = ?$ (b) $11 - 23 = ?$ (c) $2315 - 4823 = ?$

C2 Different makes of calculator have slightly different ways of showing a negative number.

Make a sketch of your calculator display for one of the answers in **C1**. Show clearly where the negative sign is on the display.

C3 Some calculators have a button labelled [+/–]. If yours has, try this:

Key in a number Press [+/–] Add 1 to the display ([+] [1])

What has the [+/–] button done?

Experiment with some other calculations using the [+/–] button. Describe what you find out to your teacher.

C4 Do this problem on your calculator: $^{-}2 - 3 = ?$

(If you did not get $^{-}5$ try again. Ask your teacher for help if you are still stuck.)

Now solve these:

(a) $^{-}25 - 77 = ?$ (b) $^{-}997 - 3 = ?$ (c) $^{-}51 + 83 - 100 = ?$

C5 The planet Mercury has one side always facing the sun. On the other side it is always night. The sunny side is up to 610 degrees warmer than the night side. If the temperature drops to $^{-}180$°C on the night side, how high could the temperature be on the sunny side?

C6 A satellite is in orbit around Mercury. Here are the temperature changes inside the satellite (in °C) every ten minutes.

$-27 \quad -86 \quad -35 \quad +35 \quad +96 \quad +73 \quad +20$

(a) What is the overall temperature change?

(b) Does it matter in which order you entered the changes into your calculator?

C7 Use your calculator to help you investigate these number pattens.

$1 - (2 \times 2) + 3 = 0$
$2 - (2 \times 3) + 4 =$
$3 - (2 \times 4) + 5 =$
⋮

$5 + 3 = 4 \times 2$
$3 + 1 = 4 \times 1$
$1 + {}^{-}1 = 4 \times 0$
${}^{-}1 + {}^{-}3 =$
⋮

Divide and rule – some talking points

There are at least four ways of working out the calculation 24 ÷ 8.

1 Try to match each of these pieces of working to what a person above is thinking. (Some may have more than one answer.)

(a) $8 \times 2 = 16$
$8 \times 3 = 24$

(b)
$$\begin{array}{r} 24 \\ -8 \\ \hline 16 \\ -8 \\ \hline 8 \\ -8 \\ \hline 0 \end{array}$$
✓ ✓ ✓

(c) $24 = 10 + 10 + 4$
$= 8 + 2 + 8 + 2 + 4$
$= 8 + 8 + 8$

(d) $24 = 3 + 3 + 3 + 3 + 3 + 3 + 3 + 3$

2 Three friends share out equally £24 they earned by cleaning cars. Which of these give how much they should each receive?

- 24 ÷ 3
- $\dfrac{24}{3}$
- 3 ÷ 24
- 24 / 3
- Twenty-four pounds divided by three.
- 24 $\overline{)3}$
- 3 $\overline{)24}$

3 Check that this works:

To divide a number by four, first divide it by two and then divide the answer by two.

Can you find some other division 'short-cuts' like this one yourselves?

4 Here are some different ways of working out 96 ÷ 4 without a calculator.

$$80 + 16 = 96$$
$$96 \div 4 = (80 \div 4) + (16 \div 4) = 20 + 4 = 24$$

$$20 \times 4 = 80$$
$$96 - 80 = 16 = 4 \times 4$$
$$\text{So } 96 = 24 \times 4$$

$$4 \times 10 = 40$$
$$4 \times 20 = 80$$
$$4 \times 25 = 100$$
$$4 \times 24 = 100 - 4 = 96$$
$$96 \div 4 = 24$$

$$\begin{array}{r} 24 \\ 4\,\overline{)\,9^{1}6} \end{array}$$

Work with a partner through each of the methods.

When you are sure how each works, calculate these:

(a) 84 ÷ 6 (b) 76 ÷ 4 (c) 96 ÷ 8

Check your answers by using at least two methods for each question.

5 Is Mark right?
Discuss it with a partner.
Tell your teacher what you have both decided.

6 Look at the division 32 ÷ 4.
What happens to the answer if you:

(a) double the left-hand side (b) double both sides

(c) halve the left-hand side (d) half the right-hand side?

Experiment with some other divisions.
What do you notice? Tell your teacher what you have found.

Review: chapter 1

1 Copy this pattern onto triangular spotty paper.
It is made up from two hexagons.

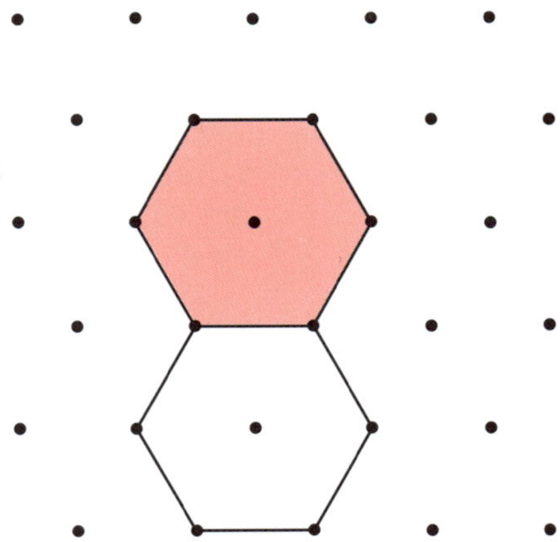

(a) Draw and shade in another hexagon with at least one side touching so that the new pattern has:

(i) reflection symmetry

(ii) no reflection symmetry.

(b) How many other patterns can you draw from three hexagons which have reflection symmetry?

2 (a) Which of these have reflection symmetry?

A

B

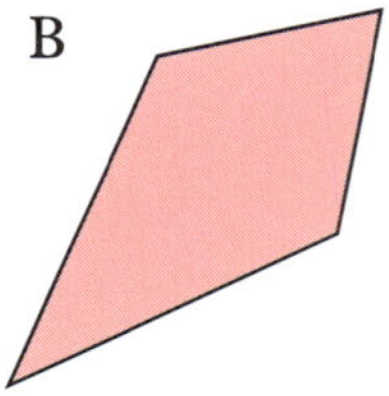

C

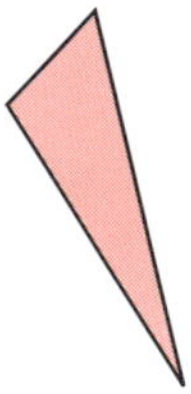

(b) Sketch the shapes which do have reflection symmetry.
Draw dotted lines for their lines of reflection symmetry.

3 Copy this carefully on to squared paper.
Shade just one square to give the pattern a line of reflection symmetry.
Draw a dotted line to show this line of symmetry.

4 Sketch a pentagon which has just one line of reflection symmetry.

5 You will need 25 multilink cubes. From these make five different solid shapes each with five cubes. Each of the solids should have a plane of symmetry. Show your solid models to your teacher.

Review: chapter 2

1 This flow diagram sorts imaginary insects into five groups.

(a) Try to work out what questions should be asked in the decision boxes.

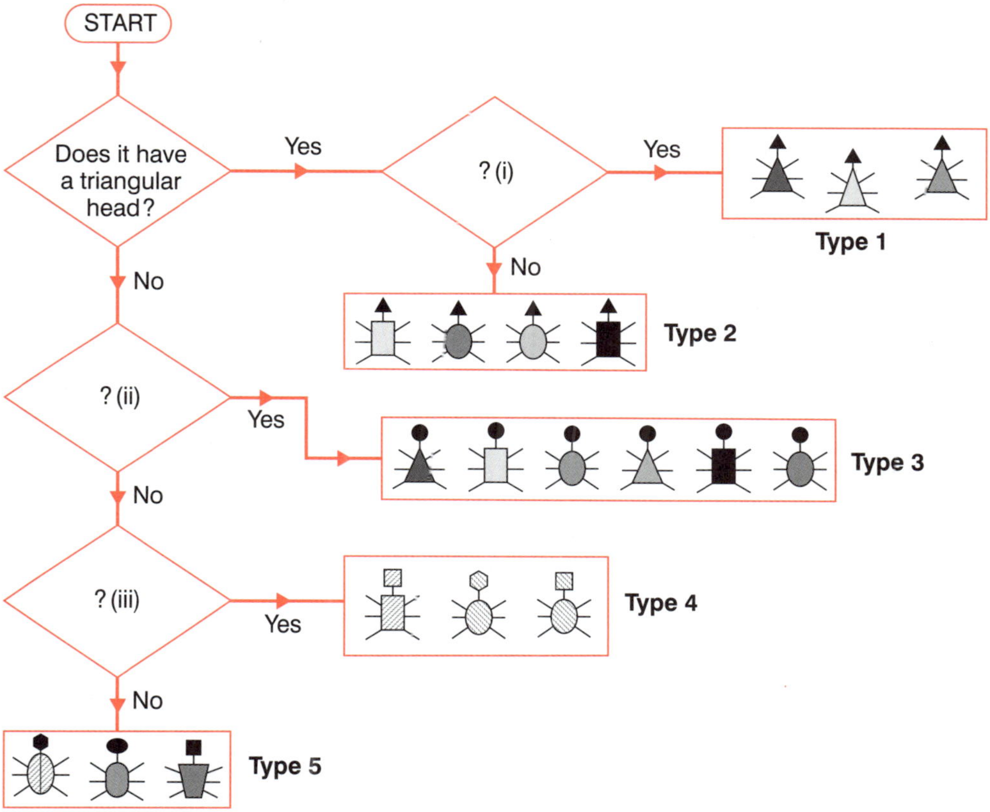

(b) Use the flow diagram to say which type these insects are.

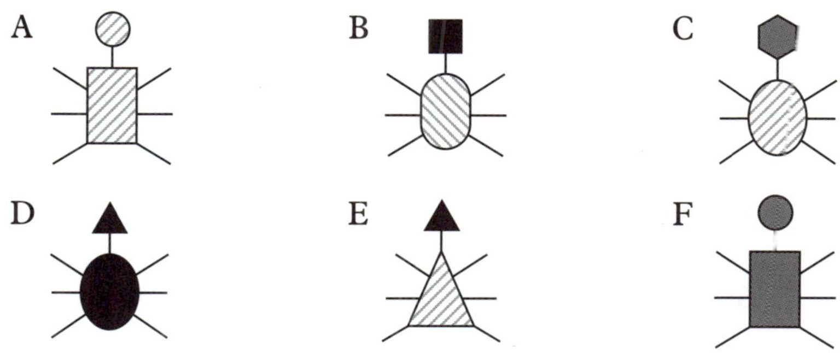

(c) Try to describe in your own words what a type 2 insect looks like.

2 Use the flow diagram below to find the names of these triangles.

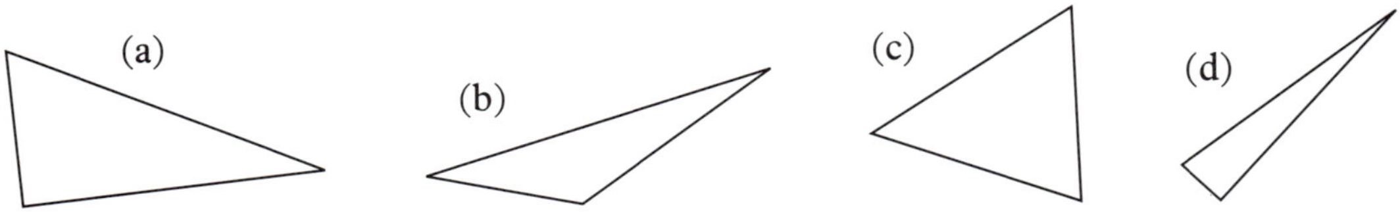

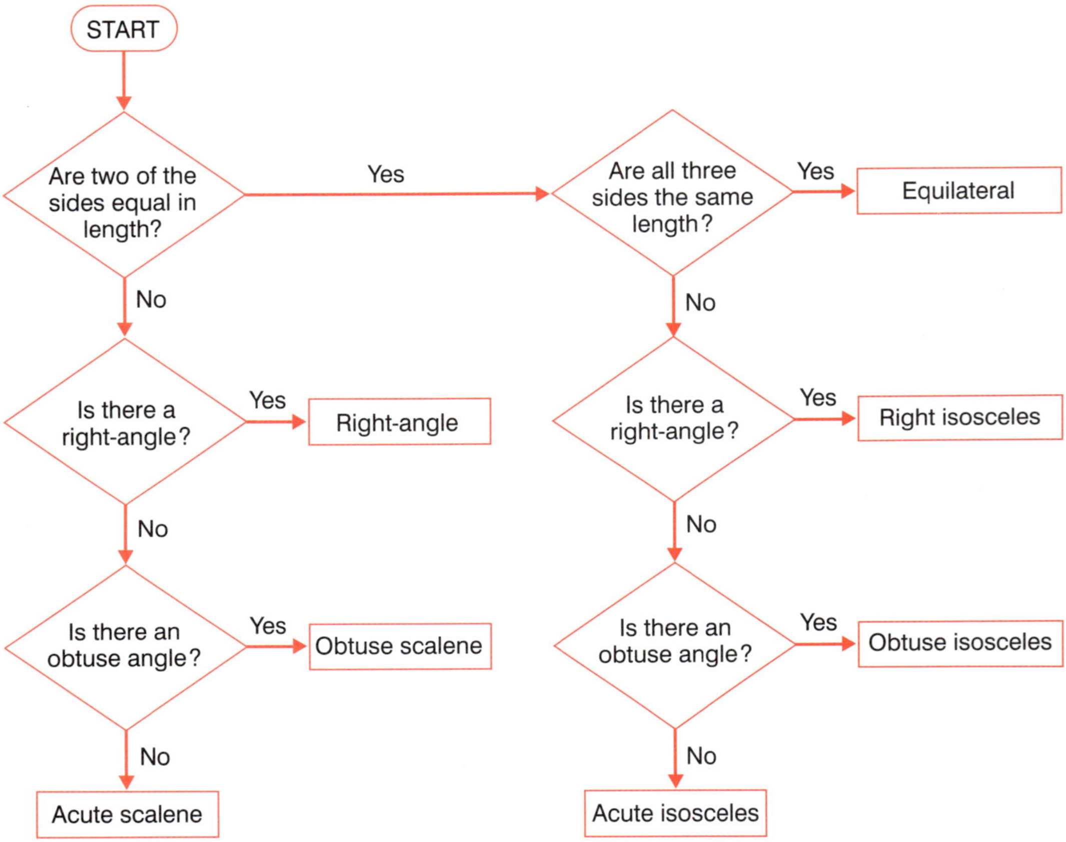

Review: chapter 3

1 This method for multiplying was used by the Ancient Egyptians. The example shows the calculation 13×9. Find out how it works. [**Hint**. It has been said that the Egyptians managed all their multiplications by doubling.]

1	9		
~~2~~	~~18~~	$2 \times 1 = 2$,	$2 \times 9 = 18$
4	36	$2 \times 2 = 4$,	$2 \times 18 = 36$
8	72	$2 \times 4 = 8$,	$2 \times 36 = 72$
13	117	$1 + 4 + 8 = 13$,	$9 + 36 + 72 = 117$

(a) Write down the calculation and answer which this working shows.

1	15
~~2~~	~~30~~
~~4~~	~~60~~
8	120
9	135

(b) Use the Egyptian method to work out these:

(i) 7×15 (ii) 11×13 (iii) 12×17

Check your answer with a calculator.

2 There are three different ways of showing 16 dots in a block. The 8×2 and 2×8 count as the same shape block.

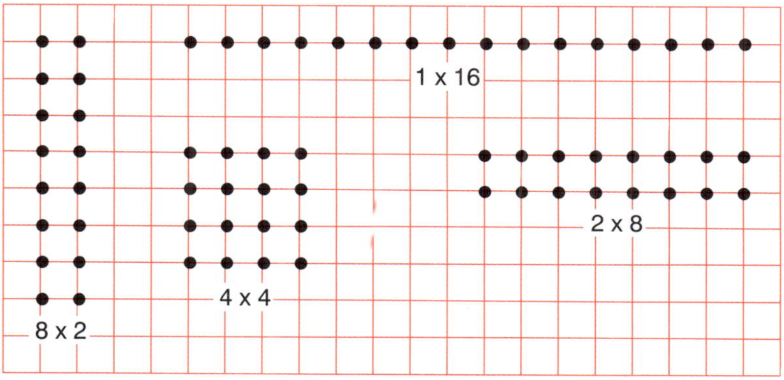

How many different ways can you find of showing

(a) 12 dots (b) 20 dots

Challenge

How many ways can you find of showing 100 in a block of dots?

3 What multiplication has this person just worked out?

$37 + 37 + 37 + 37 + 37 + 37 + 37 + 37 + 37 = 333$

4 Using just these 'facts' and some subtractions and additions, find the answers to the questions underneath.

$7 \times 57 = 399$ $4 \times 57 = 228$ $9 \times 57 = 513$

(a) 8×57 (b) 11×57 (c) 13×57 (d) 16×57

5 Work out the answers to these problems without using a calculator.

(a) Kate buys seven 19p stamps. How much will they cost?

(b)

A coach holds 54 passengers. How many will 8 coaches hold?

(c) There are 24 lorries in a yard.
Each lorry has 12 wheels.
How many wheels are there altogether?

(d) There is a big athletics match in Glasgow.
Twelve trains go from Dundee to Glasgow.
Each train has 7 carriages and there are 72 people in each carriage.

(i) How many people are there in a train?

(ii) How many carriages go from Dundee to Glasgow?

(e) A car transporter will carry 9 cars.
It needs 17 of these to clear a factory's storage space.
How many cars are in store?

(f) A swimming pool is 14 metres long. Ajit swims 5 lengths of the pool.
How many metres does he swim?

(g) An office building is 56 floors tall. The distance between two floors is about 3 metres. How tall is the office building?

6 (a) Make up a problem which fits this multiplication.

47×9

(b) Calculate the answer to your problem.

Review: chapter 4

Memory jogger

Here are the daily temperatures at midday for a week in June in Paris.

18°C, 17°C, 17°C, 23°C, 22°C, 20°C, 23°C

The **mean** temperature is (18 + 17 + 17 + 23 + 22 + 20 + 23) ÷ 7

= 140 ÷ 7

= 20°C

The temperature **range** is 23 − 17 = 6°C

The **median** (or middle temperature) is the middle one when the temperatures are put in order of size.

17 17 18 20 22 23 23

It is 20°C.

1 Here are the sales of crisps from a school shop.

Day	Mon	Tues	Wed	Thu	Fri
Packets of crisps sold	28	31	27	33	41

What was the mean number of packets sold a day?

2 Five boys and 8 girls sold programmes for a school fair. Here are the numbers of programmes they sold.

Boys	
Masood	18
Chris	6
Jason	12
Jimmy	25
Darren	19

Girls	
Becky	17
Isha	24
Amy	25
Tracy	3
Karen	10
Kathy	18
Selina	13
Janet	26

(a) What was the mean number sold by the girls?

(b) What was the mean number sold by the boys?

(c) Which group did better? Give a reason for your answer.

3 In a 5-a-side competition the mean age of a team must be under 16 years.

(a) Here are the ages of one of the teams.

16 17 16 17 15

Will the rules allow this team to play?

(b) Here are the ages of 10 players.

14 18 16 14 16 15 17 16 15 14

How would you make up two teams from these players who would be allowed to play in the competition?

4 Two seaside resorts have published their yearly rainfall figures, in centimetres, for the last nine years.

Northend	58	53	51	65	57	50	52	53	56
Whitepool	60	52	57	52	40	51	64	58	52

The tourist office at Northend says that

Are they right? Give all the reasons (and figures) you can to back up your answer.

5 Three firms make crisps. Each packet should weigh 100 g. Here are the results of weighing 7 packets of crisps from each firm. The results are shown on these number lines.

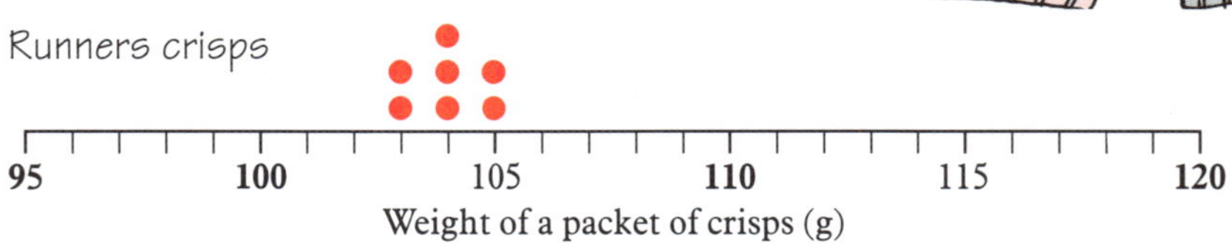

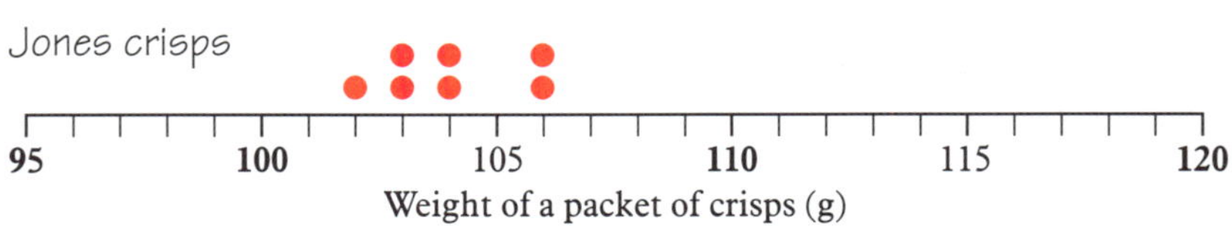

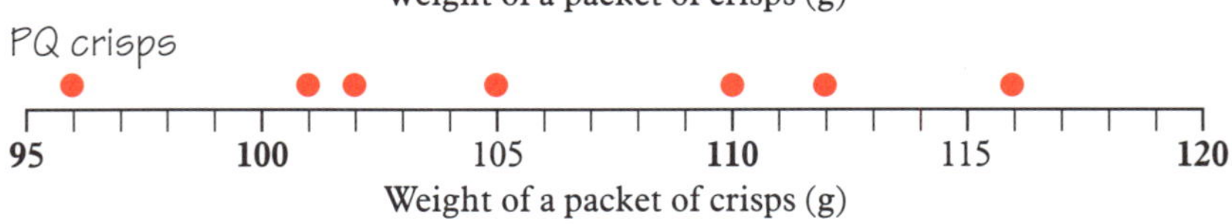

(a) Which brand had the lightest bag?

(b) Which brand had the heaviest bag?

(c) Find the range of weights for each brand.

(d) Find the mean weight for each brand.

(e) The median weight for Runners crisps was 104 g.
What were the median weights for the other brands?

(f) Which brand would you choose to buy?
Give reasons for your answers to (c), (d) and (e).

Review: chapter 5

1 Look at this sequence of shapes.

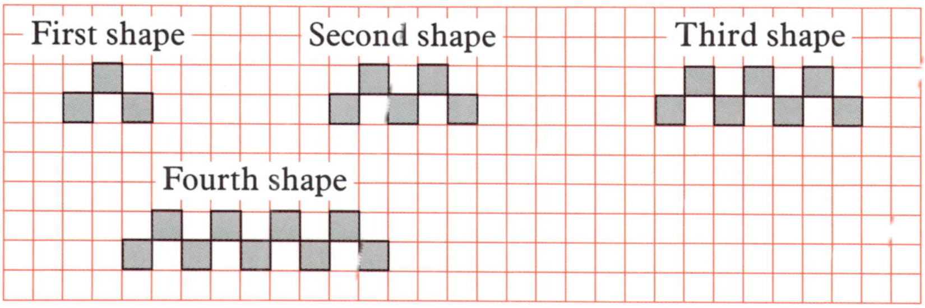

(a) Describe how to make the fifth shape.

(b) Make a list of the numbers of squares in the first four shapes.

(c) What is the rule for the numbers of squares in this sequence?

2 Look carefully at these shapes.
See how the sequence of shapes is built up.

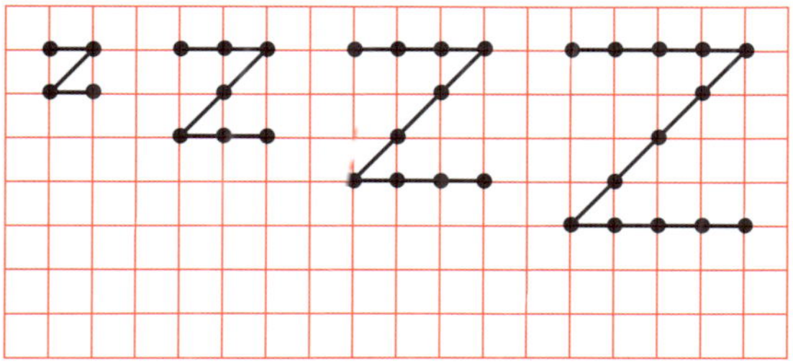

(a) Describe how to make the fifth shape in the sequence.

(b) Make a list of the number of spots in the first five shapes.

(c) How many spots would you expect to be in the sixth shape?

(d) Check your answer to (c) by drawing the sixth shape.

3 Investigate the sequence of spots made from this pattern of shapes.
Make a note of anything interesting you notice.

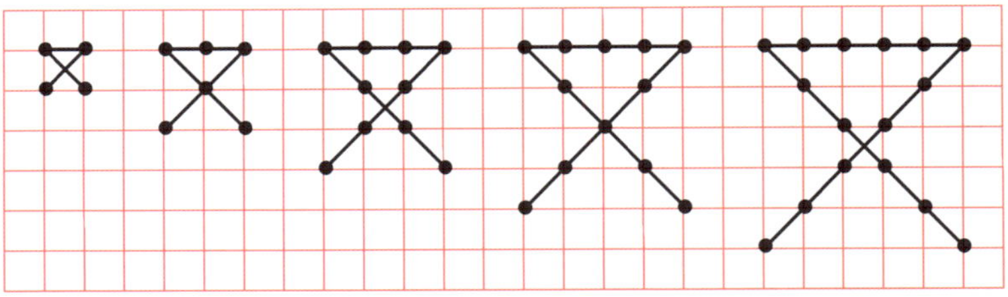

4 Which of the number machine chains below will fit this pattern?

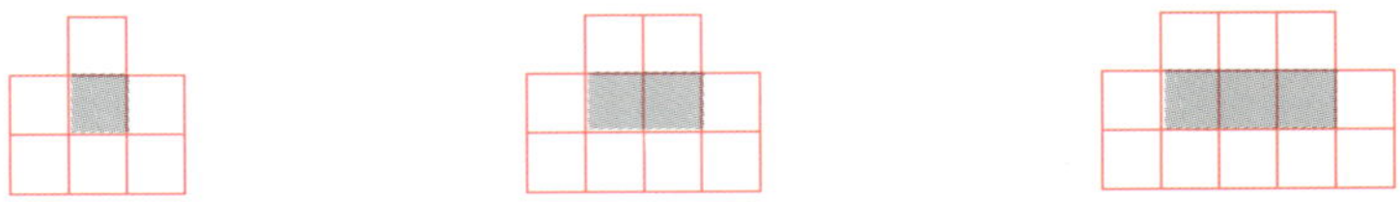

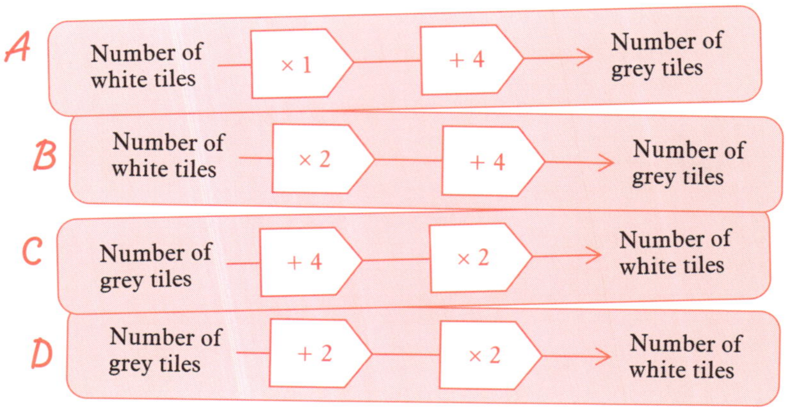

5 (a) Describe in your own words how the pattern below is built up.

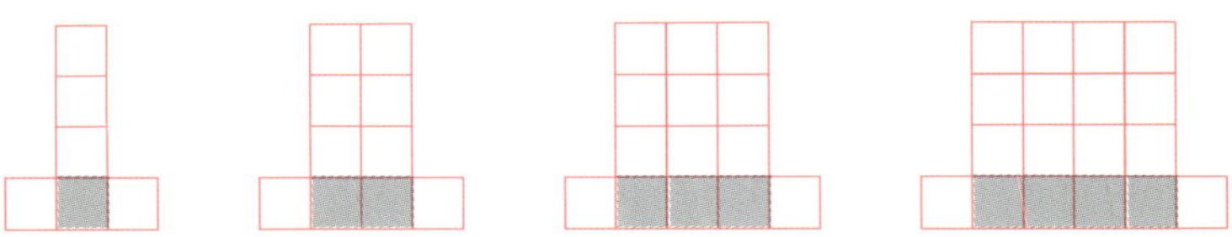

(b) Try to find a number machine chain which will give the number of white tiles for a number of grey tiles. Check that your machine chain works by drawing some more patterns.

6 Draw some sequences of tile patterns which fit this machine chain.

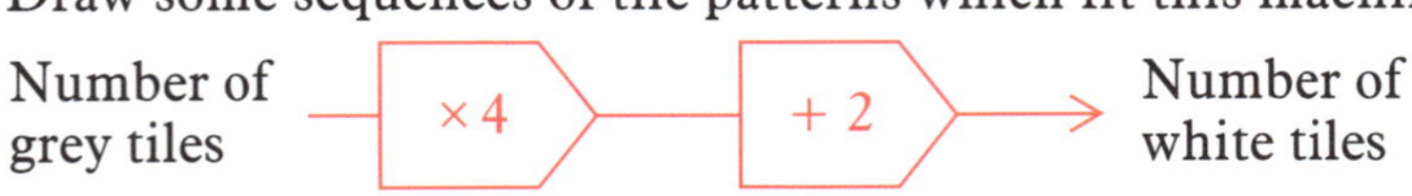

7 Draw some sequences of tiling patterns for yourselves.
For each sequence try to write down how it is built up.

Try to write the number machine chain rule connecting the numbers of differently coloured tiles (like the one in question 6).

Review: chapter 6

1 When 419 is divided by two the remainder is one.
The remainder is two when you divide by three.

What's the remainder when you divide 419 by 4?
What do you notice? Try dividing by some other numbers.

What remainders do you find when you divide 2519 by 2, 3, 4, ...?
Investigate for yourself.

2 Read this through carefully.

593 $\rightarrow 5 + 9 + 3 = 17 \rightarrow 1 + 7 =$ **8** You keep on adding the digits.

When you divide 593 by 9, the remainder is also 8!

Does this method always give you the remainder when you divide a number by nine? Experiment for yourselves.

3 Rik is playing around with the number 2520 on his calculator.
He does these divisions.

$2520 \div 2 = 1260$
$2520 \div 3 = 840$
$2520 \div 4 = 630$

Divide 2520 by 5, and by 6. What happens?
Try dividing by some larger numbers.

4 One of the largest buses built could seat 69 passengers.
It was built in Belgium.

How many of these buses would be needed to carry 1000 passengers?

5 The world's largest dumper truck can carry up to 540 tons each load.
How many loads would it take this dumper truck to move 10 000 tons?

6 Make up a problem which involves a division where you need to round the answer down.

Review: chapter 7

1 Here are six different groups of people waiting at a bus stop.

You are doing a survey on one of the topics below.
Which, if any, of the queues would you ask?
Give a reason for your answer.

(a) Where would be a good place to have a new public car park?

(b) Is there a need for a local play group for the under fives?

(c) Is the local bus service good enough?

(d) What is your favourite chart single at the moment?

(e) Where should a day centre for the elderly be located?

(f) Does the area need a summer play scheme?

2 Working with a partner, design a questionnaire for one of these.
Make a short note of when and where you would do the survey.

(a) What should the school tuck shop stock?

(b) Are school dinners popular? If not, then why not?

(c) Which CDs should be stocked in a music library?

(d) An idea of your own!

Show your teacher when you have finished.
You may be able to carry out the survey for yourselves.

3 The local council has just completed a survey on how many people live in the houses on Railway Street.

Here are their results.

Number of occupants in houses on Railway Street

5	10	6	2	6	4
0	4	8	7	8	8
5	7	3	6	6	6
2	5	4	0	4	4

(a) How many people live on Railway Street?

(b) How many houses are there on Railway Street?

(c) What is the mean number of people living in a house?

Which of these statements are **definitely** true from the survey results?

(d) There are two empty houses in Railway Street.

(e) There is no one in Railway Street living alone.

(f) Some of the houses are overcrowded.

(g) Over half the houses have five or more people living in them.

(h) Most of the people on Railway Street are poor.

(i) The mean number of people living in a house is five.
So if Railway Street was to be re-built it should be with five-bedroom houses.

Review: chapter 8

1 Heights of land are usually measured from sea-level. For example, Snowdon is 1085 metres above sea-level and the Dead Sea is 393 metres below sea-level.

What is the difference in height between the Dead Sea and Snowdon?

2 Write down the calculations needed to answer these.

(a) What is the difference in temperature between $^{-}2\,°C$ and $3\,°C$?

(b) Negative six minus two

(c) Five minus ten

3 On a warm day the temperature is about 20 °C. Estimate the temperature outside today. Check your answer if possible.

4 The temperature in outer space is about $^{-}270\,°C$.
How much warmer than this is it on a warm day?

5

On the Moon, when the sun is overhead the temperature is 117 °C.
But at night on the Moon the temperature sinks to $^{-}163\,°C$.
By how much does the temperature drop at night?

6 The tallest mountain in the world is the White Mountain in Hawaii. Its base is 6000 metres below sea-level and the peak is 4205 metres above sea-level. How tall is the White Mountain?

7 Write down a situation which fits this calculation.

$^{-}10 - 5 = {}^{-}15$